Praise for Marc Bekoff's Strolling with our Kin.

"Embracing a new and urgent environmental ethic, Marc Bekoff has made a considerable contribution to the critical mass of thinking about ecology and animal rights for the 21st century. This book will change your life, and that of your children, and their children. And, as books go, it costs less than a day at Disneyland."

— Michael Tobias, Author / editor of twenty-five books (including *Deep Ecology* and *World War III: Population and the Biosphere at the end of the Millennium*) and Filmmaker (more than one hundred films including the award-winning series *Voice of the Planet* and *Kids and Animals: A Healthy Partnership*)

"It's not every day that a world-class scientist tries to explain his ideas to young people. It's rarer still when a scientist has such an important message. Marc Bekoff's *Strolling with Our Kin* not only helps us to understand nature and animals but also shows us how to love them."

— Dale Jamieson, Henry R. Luce Professor in Human Dimensions of Global Change, Carleton College

"Marc Bekoff is the wisest scientist I know for he is the only expert who truly loves animals in the way that children are able to love animals, with all his heart. Listen to him. Read this book, give it to friends, celebrate this wonderful event."

— Jeffrey Moussaieff Masson, author of *When Elephants Weep* and *The Emperor's Embrace*

"Marc Bekoff, a biologist and one of the foremost cognitive ethologists of our time, has succeeded in writing a book that will introduce children to the field of animal ethics in a most positive way. Although Bekoff tackles complicated issues, he does so in a manner that is easily within the grasp of any young reader. This book should be a starting point for helping children to formulate an ethical approach to our relationship with the nonhuman kin with whom we share our planet."

— Gary L. Francione, Professor of Law and Katzenbach Scholar of Law and Philosophy, Rutgers University School of Law—Newark

STROLLING WITH OUR KIN

Speaking for and Respecting
Voiceless Animals

Marc Bekoff

2000

American Anti-Vivisection Society
801 Old York Road, Suite 204
Jenkintown, Pennsylvania 19046

Distributed by
Lantern Books
a division of Booklight Inc.
One Union Square West
Suite 201
New York, New York 10003-3303
www.lanternbooks.com

ISBN: 1-8816-9902-1

Dedication

For all children everywhere. May you drench your
interactions with our animal kin with respect, compassion,
spirit, humility, and love. Keep hope even when things
seem grim. Don't let go of your dreams. There are
many better tomorrows...

FOREWORD BY JANE GOODALL

FOREWORD BY JANE GOODALL

Whether an individual respects, ignores, or harms different kinds of animals depends, to a large extent, on the kind of environment in which the child grew up, especially the attitudes of family or friends. I was fortunate: when my mother found I'd taken a whole handful of worms to bed with me (I was ten months old) she did not throw them out, but quietly told me they would die without earth, so I toddled with them back into the little bit of garden outside our London apartment. She encouraged my interest in animals. We had a dog and, throughout my childhood, a series of cats, a variety of guinea pigs, hamsters, and so on which my sister and I were absolutely responsible for – but not allowed to forget. Ever. We were brought up to hate caged things – our canary and subsequently our budgerigar, had cage doors left open, and flew freely. Our rabbits had the run of the house, the hamster nested in the back of the sofa and our guinea pigs were regularly taken from their runs and walked, on harnesses, around the streets.

Also, we were encouraged both to explore and interact with the natural world, and to learn about it from the books with which my childhood was surrounded. The most precious of these was a large volume The Miracle of Life, which had been delivered free as a reward for collecting an astronomical number of coupons from packets of some kind of food – cereal probably. That book, with its myriad of illustrations, dealt with topics as diverse as "Many tongues for many purposes" and the history of medicine. It was written for adults but it was my passion, and helped to shape my respect for life and wonder in nature.

Since those far off days, before the second world war, much has changed in the world. There always was abuse of animals but we are more aware of it today thanks to the media, and to the tireless efforts of those who seek to protect creatures of all sorts, from lobsters to chimpanzees, and the environments in which they live. The horror of factory farming is new. The extraordinary explosions of human populations world wide has caused an ever increasing hostility between "man" and "beast" as they compete for dwindling resources – and the natural world is losing out. The grim inner-city areas and the poverty that exists even in the most affluent countries increasingly alienates children from nature.

There is a new need for information that will encourage young people to understand the natural world and their relationships to it. A new need to teach children in school about the way their societies treat animals. And a new need to provide our youth with opportunities that foster respect for all life and an empathy with the animal beings with whom we, human beings, share the planet. This is why I developed our Roots & Shoots program that requires groups of young people to participate in projects that benefit the environment, animals, and the human community in their area. A program with two major messages: (a) that every individual matters and can make a difference, every day; and (b) "Only when we understand can we care; Only when we care shall we help; Only if we help will all be saved." We must encourage our children, empower them to help us save the world that will, so soon, be theirs. There is a vast amount of information about issues of animal abuse and conservation in a vast number of books, magazines, and the children's sections of major conservation and animal welfare organizations. And Marc Bekoff has pulled the issues together and written about them with clarity and conviction.

Strolling With Our Kin is a reference book which can be helpful to children, parents, and teachers alike. It tackles all aspects of our relationships with animals, both in the wild and in the home. It deals with ethical issues about the way we exploit animals, and think about them - not only in relation to ourselves, but as beings in their own right with their own needs, to be celebrated and respected.

When children have access to good information that enables them to have a good understanding of these issues, their logic is impeccable. My grandson, aged five, has grown up in Tanzania surrounded by people who look upon fishing as a way of life. He has also been encouraged to respect animals. A few months ago, after spending much time with one of our Roots & Shoots volunteers, he asked his father where the fish on his plate came from. Was it killed for him? In that case, please, he did not want to eat any more fish. Quite independently, his three year old sister asked similar questions about the chicken served for dinner - with the same result. When I checked with the volunteer, she said she had not advocated vegetarianism - though she is one herself - as am I, though not my son. But she had talked a lot about how animals feel and figure things out.

If children know about the terrible cruelty that may result from the pet trade they will understand why it is a mistake to buy an exotic animal. If they are exposed to what often goes on behind the scenes in seemingly good zoos, they will view the animals there differently - they will be asking questions about issues they were unaware of before.

If knowledge of animals leads to respect and concern for their welfare, the reverse can be true also. Every time a reluctant child is forced to dissect a once living creature in class, it will be that much easier the next time. There are other ways to learn respect for living forms, to wonder at their workings and their variety - ways that are not destructive of life.

Strolling with Our Kin will encourage children to think for themselves, to question the status quo. And it will provide parents and teachers with invaluable information about subjects about which they may know little, such as animal intelligence, their capacity to feel pain, their use in pharmaceutical testing, medical experimentation, intensive farming, training for entertainment - and so on. The subjects covered are fairly exhaustive.

Above all, *Strolling With Our Kin* will encourage the natural curiosity of children in their world, foster a sense of wonder and delight, and a corresponding sense of concern. And children who are kind to and respectful of animals are likely to show more understanding in their relations with other humans as they grow up.

I hope this book will soon be available in all libraries and on the shelves in many homes. Certainly I shall be recommending it to all 1500 of our Roots & Shoots groups in North America, and in other parts of the world as well.

CHAPTER 1

THE ABC'S OF ANIMAL
PROTECTION AND COMPASSION

Living in harmony with our kin in the more-than-human-world

Always **Be** Caring and Sharing - that is what the **ABC'S** of animal protection and compassion are all about. And that is also what this primer is about. To make things simple, I will use the word "animal" to refer to non-human animal beings, recognizing of course that humans are indeed animals, as are our closest kin, the Great Apes such as chimpanzees and gorillas, and also dogs, cats, rats, snakes, lizards, fish, bees, ants, and termites.

Nowadays, questions about animal-human relationships are of great importance to more and more people. In Universities, researchers in many areas - philosophy, biology, psychology, anthropology, history, sociology, and law - are all working together to provide answers for the many and complex questions concerning animal-human interactions. As well, many people who do not work in Universities are very interested in how humans relate to other animals. They spend a great deal of time trying to make the lives of other animals better for the animals themselves.

There are numerous issues that need to be considered when discussing animal-human relationships. I can present only a few here. Fortunately, many people believe that the lives of other animals are important, and they try very hard to make animals' lives the best they can be. They want other animals to be happy, well taken care of, and respected. While there are other people who believe that we can do whatever we want to animals because we are humans - we are the "master" species – there are others who believe that we should never interfere negatively in the lives of other animals, especially on purpose. These people believe additionally that we are not the "master" species but one among many.

Moderation and consistency

Most people take a moderate position on animal use by humans. They allow some but not all animal use. They feel all right about the use of some animals rather than others. For these people, all animals are not equal. They often find it difficult to be consistent and objective. Maybe it would be all right to use chimpanzees to save their own mother's or child's life, but not the life of someone else's mother or child. Perhaps it is fine to have a fish in an aquarium or a bird in a cage but not a gorilla in a zoo. As Lisa Mighetto emphasizes in her book *Wild Animals and American Environmental Ethics*, "Those who complain of the 'inconsistencies' of animal lovers understand neither the complexity of attitudes nor how rapidly they have developed." Even with our inconsistencies and contradictions when dealing with the difficult issues centering on animal protection, we have come a long way in dealing with many, but not all, of the problems. But we should not be complacent, for there still are far too many animals suffering at the hands of humans, and much work still needs to be done. I am not trying to criticize these people, for the issues are very difficult. But some degree of consistency and perhaps strong guidelines are necessary to guide us, so that we can lessen the pain and suffering that humans cause to other animals every second of every day.

Why all the concern about how animals are treated by humans; why do some people spend a large portion of their lives studying animal-human interactions rather than playing games, going on vacations, or trying to learn about other interesting aspects of the wondrous world in which we all live together? When many people sit back and look around at the world they realize that they are too far removed from the other animals - and even too far removed from plants, rocks, and streams - with whom they share the planet Earth. This distance has made the world a mess - there is horrible pollution, too much disease, too much stress; there are too many cars and too many abused animals whose lives have been ruined. Many people are coming to realize that they are a part of the rest of nature and not *apart* from nature. *No one is outside nature.* These people want to do something for all beings with whom they share the precious and limited resources on Earth. What better place to start than with the other animals - our kin - with whom we share space and with whom we can actually "talk?"

Why I am writing this book:
The view from within biology

My early scientific training as an undergraduate and a beginning graduate student was grounded in what the philosopher Bernard Rollin calls the "common sense of science," in which science is viewed as a fact-gathering value-free activity. Of course, science is not value-free - we all have a point of view - but it took some time for me to come to this realization because of the heavy indoctrination and arrogance concerning the need for scientific objectivity. In supposedly objective science, animals are not subjects but rather objects that should not be named, and objects with which close bonding is frowned upon. However, naming and bonding with the animals I study is one way for me to respect them. Although some people believe that naming animals is a bad idea because named animals will be treated differently - usually less objectively than numbered animals - others believe just the opposite, that naming animals is a good idea. As Christopher Manes notes about many Western cultures, "If the world of our meaningful relationships is measured by the things we call by name, then our universe of meaning is rapidly shrinking. No culture has dispersed personal names as miserly as ours . . . officially limiting personality to humans . . . [and] animals have become increasingly nameless. Some*thing* not some*body*." In her book *Reason for Hope*, Jane Goodall, the world-famous expert on chimpanzee behavior and tireless crusader for generating human respect for animal lives, notes that early in her career she learned that naming animals and describing their personalities was taboo in science, but because she had not been to University she did not know this. She "thought it was silly and paid no attention." Dr. Goodall opposed reductionistic, mechanistic science early in her career, as she does now, and her bold efforts have had much influence on developing scientists' views of animals as thinking and feeling beings.

I am a biologist, a lover of the diverse and wondrous life on this splendid planet. As a scientist who has been lucky enough to have studied social behavior in coyotes in the Grand Teton National Park in Jackson, Wyoming, the development of behavior in Adélie penguins in Antarctica near the South Pole, and social behavior in various birds living near my home in the Rocky Mountains of Colorado, I have learned a lot about these amazing animals and many others.

I am very concerned about what humans are doing to other animals and to the planet in general. While some of my views may make it seem as if I want to stop *all* animal research, including my own, and the use of all animals everywhere, this is not so. I am just not very happy with what is happening to the wonderful animals with whom I am privileged to live and share the Earth. Are you?

In this short book I will discuss some broad and interrelated topics and raise numerous questions about how animals and humans interact and how animals are used for mostly human benefits. Each topic makes a number of points. All topics are related to the main issue - *the choices we make when we interact with other animals with whom we are privileged to share the planet.*

It will become clear that the nature of animal-human encounters, how animals are viewed and treated, has large - often enormous and irreversible - impacts on the many different environments in which we live. As you read this primer you will discover how closely related questions are when dealing with whether or not animals are as valuable as humans, whether or not animals have rights, whether or not animals are conscious, whether or not animals feel pain and suffer, and whether or not individual animals count more than entire species. The answers that are given for these and other questions greatly influence how humans view other animals and interact with them.

I have also compiled a list of resources to help you learn more about these and other topics, because some of the questions that need to be considered are difficult, and it is helpful to see what others have to say about them. There is always something to be learned from others' views.

I hope this primer will appeal to people of all ages and in different cultures, for the issues I discuss and the questions I pose have few if any boundaries with respect to age, culture, and time. I am also hoping that this primer will serve to generate many more questions than those I raise, and that perhaps some people will be able to read it together and discuss questions as they arise. Siblings can read it to one another, parents, teachers, and other adults can read it along with children. Young children are very interested in animals. Their initial contacts usually are friendly and indicate they do not recognize many large differences between themselves and other animals. When my nephew Aaron was two years old, he knelt, went nose-to-nose with a worm, and said "hello." At ten, Aaron still is naturally attracted to animals.

Much of what children come to believe about and feel for animals is learned and influenced by early environments. Older children and adults can play major roles in developing children's attitudes that have long-lasting effects. Jane Goodall's Foreword says this all very nicely.

Why you are so important

Why are you so important? What difference can you make? Why should you care about other animals and the environment? It is very easy to answer this question. We all live on this planet and inherit the Earth that others leave behind. By thinking about these issues it is likely that you will become more closely attached to the other living organisms and inanimate objects around you. Animals count, trees count, and rocks count. But all too often we live as if future generations do not count. So what if people in the future inherit the messes we leave? We need to share our lessons with each other, for what others learn will influence how they think and act.

It is important to encourage others to explore how animals live, important for all of us to want animals' lives to be the best they can be, and important for you to ask questions about how humans treat other animals. It is important for all of us to know that hamburgers were once cows, that the bacon on a bacon, lettuce, and tomato sandwich was once a pig, and that cows, pigs, chickens, and fish are social animals and have families just like humans. Both the cow and the pig were once someone's child, brother, or sister. They had lives that were ended so that people could eat them. They were removed from their mothers or families, housed in horrible conditions, shipped to commercial food processing plants, and killed, suffering all the way. While this description sounds awful, and it could be colored by using other words to make it less offensive, this really is what happens to cows, pigs, and other animals who become human meals. If we do not tell it like it is, important messages are lost.

Teaching and practicing tolerance are good habits to incorporate into all of our lives. *We need to develop and to live an ethic of caring and sharing, so that all animals are respected for the individuals they are.* Perhaps the best way to state it is that we need to recognize that we are privileged to live on such a wonderfully diverse planet that is full of incredible and bountiful beauty. So that our children, our children's children, and their children in turn can fully enjoy the beauty and grandeur that nature offers, everyone must give very serious attention to how animals are viewed and treated. *We are so lucky to have so many other animals as our friends.*

Habitat loss and planetary biodiversity

Globally and locally, within small communities, there is much interest in the many and difficult questions concerning how humans interact with and treat animals. Of course, as we will see, global and local issues are closely related to one another. How and why humans and animals interact in nature, in industry, in zoos, in wildlife theme parks and aquariums, and in research laboratories are very important and controversial topics all over the world.

Globally, populations of humans are growing rapidly, and many populations of wild animals and plants continue to lose their battle with humans. Global biodiversity - the number of different species that inhabit our planet - is rapidly, and perhaps irreversibly, dwindling. Recently, there was a front-page story in the *New York Times* titled "It is Kenya's Farmers vs. Wildlife, and the Animals are Losing." Indeed, fifty-eight percent of the animals in Kenya's Tsavo region, about 106,000 large mammals, vanished between 1973 and 1993. Problems such as Kenya's concerning farming, tourism, human interests and needs, and the fate of wild animals are issues world-wide. They demand close attention now because of the enormous uncontrolled growth in the number of humans all over the planet, the decline of habitat where animals can live (in Kenya, it is estimated that wild lands are disappearing at a rate of two percent a year), and the increasing use of animals to meet human needs and desires.

On the global level, many researchers think that the main problem is fairly simple - there are too many people and not enough land to house and feed them. Indeed, habitat loss is considered by most conservation biologists to be the biggest threat to animal and plant life. Uncontrolled habitat loss means there will be a loss in global biodiversity. Even if humans want to reintroduce species to the wild or relocate them to suitable habitats where they would be able to thrive and survive, such places will not be available because humans continue to develop these areas, thereby making them unsuitable for either the return or the relocation of animals.

Animal use: The numbers speak for themselves

In addition to global issues concerning biodiversity, there are also local concerns that center on individual animals rather than on entire ecosystems, populations, or species of animals. Because there are so many people, the demand for animal products and for dealing with human medical needs and food requirements is rising astronomically.

The numbers tell the grim story. Gail Eisnitz, in her book *Slaughterhouse: The Shocking Story of Greed, Neglect, and Inhumane Treatment Inside the U. S. Meat Industry,* reports that in the United States alone, 93 million pigs, 37 million cattle, 2 million calves, 6 million horses, goats, and sheep, and nearly 10 billion chickens and turkeys are slaughtered for food each year.

Animals are also used in experimental research. In 1996, according to a survey by the United States Department of Agriculture, the number of animals used in experimentation totaled about 1.3 million individuals, including fifty-two thousand non-human primates, eighty-two thousand dogs, twenty-six thousand cats, 246,000 hamsters, and 339,000 rabbits. This staggering number does not include rats, mice, and birds, who make up about ninety percent of the animals used in experimental research. These three species are not given even the most basic care and housing offered to other species under the federal Animal Welfare Act. It is estimated that more than 70 million animals are used annually and that one animal dies every three seconds in American laboratories.

Government workers also kill numerous animals. For example, some people who work for the Bureau of Land Management and the United States Fish and Wildlife Service "recreationally shoot" and kill prairie dogs in order to control populations of these beautiful rodents. The Animal Damage Control (ADC) unit (now called Wildlife Services) of the United States Department of Agriculture is responsible for cruelly and indiscriminately killing hundreds of thousands of animals - varmints they call them - including coyotes (over eighty-two thousand in 1997), foxes (5,858) and mountain lions (320) in the name of control and management. In 1997 Wildlife Services killed over ninety-one thousand predators. However, only about one percent of livestock losses are due to predators, and ninety-nine percent are due to disease, exposure to bad weather, illness, starvation, dehydration, and deaths at birth. Wildlife Services is also responsible for negatively influencing populations of at least eleven endangered species.

Numerous animals are also genetically engineered or designed so that they might develop heart failure at an early age or develop resistance to various diseases (see the American Anti-Vivisection Society's (AAVS) Animalearn Fact Files, Genetic Engineering). Genetically altered food, called "Franken-food," is becoming very popular (see: www.purefood.org, www.bio-integrity.org, and www.safe-food.org). Furthermore, Bovine Growth Hormone (rBGH) is being used to increase milk production by cattle, despite its demonstrated risks to the health of cattle (increased udder infections and foot diseases) and humans (possible increased risk of breast cancer). The Humane Farming Association (www.hfa.org/) is leading a national campaign to protect consumers from the dangers of agri-chemicals and to protect farm animals from being abused for profit. Bernard Rollin has written about these issues in his book *The Frankenstein Syndrome: Ethical and Social Issues in the Genetic Engineering of Animals,* as has Michael W. Fox in *Beyond Evolution: The Genetically Altered Future of Plants, Animals, the Earth . . . and Humans.*

We also need to be deeply concerned about the well-being of our companion ("pet") animals. Michael Tobias, in his book *Voices From the Underground: For the Love of Animals,* reports that according to two surveys taken in 1994, there were about 235 million companion animals in the United States, including 60 million cats, 57 million dogs, 12.3 million rabbits, guinea pigs, hamsters, gerbils, and hedgehogs, 12 million fish tanks (no estimate of numbers of fish), 8 million birds, 7.3 million reptiles, and 7 million ferrets. About seventeen billion dollars per year is spent on supplies. While it is wonderful to think that the sheer number of animals indicates that people truly care about their companions, this is not so. Far too many animals breed, and there are numerous unwanted individuals. Many are ignored or abused when they become burdensome to their human companions. Many are also tortured "for fun." Numerous organizations, including local Humane Societies, have programs directly concerned with the well-being and fate of companion animals. The Humane Society of the United States has a program called First Strike devoted to looking at the relationship between cruelty to animals and cruelty to humans.

The more-than-human world

When human populations show explosive growth, it is other animals - entire ecosystems, populations, species, and individuals - who suffer. In the end, it is fairly simple: animals lose when human interests come into conflict with animal interests.

The problems we face in the area of animal-human interactions raise numerous questions, many of which I consider in this book. For example, how *should* humans treat other animals? There is much interest in whether or not humans *should* treat other animals in particular ways. Do we have to treat animals in certain ways? Are there right and wrong ways for humans to treat other animals? Can we do whatever we want to other animals? Do we need to respect animals' rights? Do animals even have *rights*? And if animals have rights, what does this mean? People interested in the issues centering on animal-human relationships are concerned with the *ethics* of animal-human interactions.

It is important to emphasize that for many questions about how animals should be treated by humans there are not "right" or "wrong" answers. However, there are "better" and "worse" answers. Open discussion of all sides will help us make progress. No one view can be dismissed by pretending that it does not exist. Ignoring the problems will not make them disappear. How we relate to animals is closely related to how we relate to ourselves and to other humans.

As the philosopher and story-teller David Abram reminds us, we live in a *more-than-human world*. Native Americans are proud to claim, "animals are all our relations." These are important messages because they stress the close, intimate, and reciprocal relationships that exist between animals and humans. Animals are our kin. It is important to remember that many animals give much to us. They teach us about trust, respect, and love. They are there for us without any conditions, other than that we keep them alive by feeding them, give them a place to live, and care for them when they are healthy and sick. Animals also teach us about responsibility. We need to make sure that all animals have every opportunity to enjoy themselves - to be happy and content. Compassion and respect are the least that we can give back to them.

It is also important to remember that when humans choose to use animals, the animals invariably have no say in these decisions. They cannot give their consent. Animals depend on our good will and mercy. They depend on humans to have their best interests in mind.

Speaking for voiceless animals

Humans are constantly making decisions for animals. *We are their voices.* So when we speak for them, in order for there to be balance, we need to be sure that we are taking into account their best interests. As Jane Goodall has said: "The least I can do is to speak out for the hundreds of chimpanzees who, right now, sit hunched, miserable and without hope, staring out with dead eyes from the metal prisons. They cannot speak for themselves." I cannot ask my companion dog Jethro if it is all right for me to use him in an experiment in which he will suffer and feel pain, and perhaps be killed. When humans use chimpanzees for behavioral or medical research, they do not ask them if they agree to be kept in small cages alone, be injected with viruses, have blood drawn, and then perhaps be "sacrificed" - killed - so that the psychological and physiological effects of the experiment can be studied in more detail.

Because of our position in the world, because we can freely speak and express our feelings about animals, and animals do not have much say in the matter, animals *seem* to be there for us to use in any way we choose. However, animals should not be viewed as property, resources, or disposable machines who are here for human consumption. Animals should not be treated as we treat bicycles or backpacks. It is important to remember that we do not *have* to do something just because we can do it. Just because certain activities *seem* to have worked in the past does not mean that they truly have worked. For example, many people believed that invasive experimental research on animals and eating meat were *essential* for human betterment, whereas we now know that this is not so. While we cannot undo all the mistakes people have made, there is still time - perhaps not a lot of time - to make changes that will help us and other animals along.

Unfortunately, many people are largely detached from nature and the outdoors in general. A recent survey showed that many people spend more than ninety-five percent of their lives indoors. One result of this detachment from nature and our animal kin is that animals come to be treated as if they are property - items that we own such as our bicycles or backpacks - for humans to do with what they want. However, animals are not mere resources for human consumption. That is not why they have evolved to be the splendid beings they are.

I hope that this book promotes open discussion about how we treat other animals. Open discussions can make us better people and through such discussions, we can create a better world in which to live. Sometimes it is useful to take the opposite point of view from that with which you agree, and try to defend it. You then can imagine what your opponents are thinking and how they develop their own thoughts and feelings about animal-human interactions.

Minding animals: A personal story

A good place to begin is with a few short stories about my own experiences with animals. I did not grow up in the company of many animals. From time-to-time there were some goldfish and small painted turtles with whom to share my life. I loved them and watched them do what they do.

My parents tell me that when I was very young, I "minded" animals. "Minding animals" means that I cared for them and simply assumed that they had very active brains and minds. I never doubted that they were very smart. I always asked "what do you think they're thinking - what's on their minds?" I never doubted that animals were just like us in many ways. Of course, I came to realize that animals have their own points of view, but by using human terms to describe animal emotions and behavior - they are happy, sad, angry, jealous - it made it easier to tell myself and others what might be happening in their heads. It also struck me as odd that many people who thought that animals could have negative emotions - that they could be angry, mean, or depressed and treated with antidepressants - were uncomfortable when people ascribed positive emotions to them - that they could be happy and enjoy life.

I did not have the wonderful experience of sharing my life with a companion dog until I was twenty-six years old. Moses was a large white malamute who blessed my life for only a short period of time. He was a bundle of joy and made me realize how much I had missed when I was a kid. Because I was in graduate school studying animal behavior, it was natural to include him in my circle of dearest friends. When he was two-and-a-half years old he died while being treated for a hang-nail. A simple hang-nail. The veterinarian gave him a mild sedative, and Moses had an allergic reaction to it and died suddenly. I (and the veterinarian) was devastated - how could this happen, how could I cope with his absence? It was the first time that I had experienced such a great loss, and it made me even more determined to live with and to study animals for the rest of my life. I wanted to learn more about their lives, how and why they come to mean so much to so many people, and what I could do to make their lives better for them, regardless of what I or other humans wanted or needed.

Missing mom: Candid coyotes

Here is a story that always creeps into my mind when I think about animal tales that show how "minding animals" can help us understand their lives. Years after this incident happened, it still rings clearly.

My students and I studied coyotes for seven years around Blacktail Butte, in the Grand Teton National Park, south of Jackson, Wyoming. A female who we called "mom" was a mother and a wife from the beginning of the study; in late 1980, she began leaving her family for short forays. She would take off and disappear for a few hours and then return to the pack as if nothing had happened. I wondered if her family missed her when she wandered about. It sure seemed that they did. When mom left for forays that lasted for longer and longer periods of time, often a day or two, some pack members would look at her curiously before she left - they would cock their heads to the side and squint and furrow their brows like they were asking "where are you going now?" Some of her children would even follow her for a while. When mom returned, they would greet her effusively by whining loudly, licking her muzzle, wagging their tails like windmills, and rolling over in front of her in glee. "Mom's back!" Her children and mate missed her when she was gone.

One day mom left the pack and never again returned. The pack waited impatiently for days and days. Some coyotes paced nervously about as if they were expectant parents, whereas others went off on short trips only to return empty-handed. They traveled in the direction she had gone, sniffed in places she might have visited, and howled as if calling her home. For more than a week some spark seemed to be gone. Her family missed her. I think the coyotes would have cried if they could. It was clear that coyotes, like many other animals, have deep and complicated feelings. Their behavior told it all.

After a while, life returned to normal on Blacktail Butte. Sleep, eat, play a little, hunt, defend the territory, rest, and travel. A new and unfamiliar female joined the pack, was accepted by all the coyotes, formed a partnership with the breeding dominant male, and eventually gave birth to eight babies. She was now mom and wife. But every now and again it seemed that some of the pack members still missed the original mom - maybe she was lost, maybe she would return if we went to look for her. The coyotes would sit up, look around, raise their noses to the wind, head off on short trips in the direction that mom last went, and return weary without her. It took about three or four months until these searches ended. Pack members still seemed to miss mom, but enough was enough. There were things that needed to be done that could not be put off any longer.

Rabbit punching

Here is another story of some events that changed my life. One afternoon during a graduate course in physiology, one of my professors calmly strutted into class sporting a wide grin and announced that he was going to kill a rabbit, to be used by us in a later experiment, by using a method named after the rabbit himself, namely a "rabbit punch." He broke the rabbit's neck by chopping him with the side of his hand. I was astonished and sickened by the entire spectacle. I refused to partake in the laboratory exercise and also decided that what I was doing at the time was simply wrong for me. I began to think seriously about alternatives. I enjoyed science and continue to enjoy doing scientific research, but I imagined that there were other ways of doing science that centered on incorporating respect for animals and allowing for individual differences among scientists concerning how science was conducted. I went on to another graduate program but dropped out because I did not want to kill dogs in physiology laboratories and cats in a research project. Recently I learned that the famous biologist Charles Darwin might also have left medical school after one year because he was "repulsed" by experiments on dogs. In his book *The Descent of Man*, Darwin wrote the following about someone who experimented on dogs: " . . . this man, unless he had a heart of stone, must have felt remorse to the last hour of his death."

Finally, I found myself studying the behavior of animals in a graduate program in which I could watch them, record what they did, and not have to kill them to learn about their exciting lives.

It is all right to care about animals

Now, the reason I am telling you all this is not to be preachy, or to say that mine is the only route to follow. Rather, it is to make a few points. First, while having contact with animals early in life might be important for developing empathy and feelings for them, it is not necessary. Also, it is possible for people to change their views on animals. I have changed my mind about what I can and cannot do to animals, and so have a number of my colleagues.

The famous Nobel-prize winning ethologist Konrad Lorenz was of the opinion that you should love the animals you study, that it is all right to bond with them. Some researchers do not like to form close bonds with the animals they study because they think it influences their research. However, forming bonds and respecting the animals with whom one works might make one see animals differently than if the animals were viewed as mere objects. Respecting animals for whom they are in their own worlds, and respecting animals as subjects of a life and not mere objects, can only make "science" better, or more reliable, because when we respect each animal's point of view we will stop thinking of animals from our own points of views but rather theirs.

While other animals may be different from us, this does not make them less than us. Animals have their own lives and their own points of view, and it is important to recognize this. Maybe it is easier to harm other animals if we distance ourselves from them - we are so different from other animals, we tell ourselves, that it is all right to harm them. But, this only makes matters worse for the animals and in the end, I think, for humans. People who do this get so removed from the world around them they cannot appreciate its remarkable beauty and splendor.

It is all right to care about animals. By recognizing the beauty and value of each and every life, I believe the world will become a better place and that better science will result. We are animals' guardians and spokespersons, and we owe them unconditional compassion, respect, and support, as we do people. We have control and "dominion" over other animals, but this does not mean that we have the right to exploit and dominate them. Most importantly, each and every one of us makes a difference.

Some guiding principles

I consider myself lucky and privileged to have been able to have made the intimate acquaintance of many and diverse animals, to have been touched by whom they are. I am sure that in some instances they were watching, smelling, hearing, and studying me as closely as I was observing them. Often people are not aware that they are interfering in the lives of the animals in whom they are interested. *The guiding principles for all of our interactions with animals should stress that it is a privilege to share our lives with other animals, we should respect their interests and lives at all times, and the animals' own views of the world must be given serious consideration.*

The animals' influence, that resulted from their unselfish and intimate sharing of their lives with me, is clearly reflected in my views on animal minds, animal rights, and science. These include: (1) putting respect, compassion, and admiration for other animals first and foremost; (2) taking seriously the animals' points of view; (3) erring on the animals' side when uncertain about their feeling pain or suffering; (4) recognizing that almost all of the methods that are used to study animals, even in the wild, are intrusions on their lives; (5) focusing on the importance of individuals; (6) appreciating individual variation and the diversity of the lives of different individuals in the worlds in which they live; and (7) using common sense and empathy when doing scientific research. Although I have always been concerned with animal rights, I have not always applied the same standards of conduct to my own research. I have done experiments that I would never do again: on predatory behavior in infant coyotes in which mice and chickens were provided as bait. Coyotes were allowed to chase and kill the mice and chickens and the mice and chickens could not escape. I am sorry I did this type of research, and I apologize to the animals who I allowed to be killed.

Important questions to ponder

Now, what are some questions that must be considered when talking about how humans and animals get along? Why should we care about other animals? First, I will list some questions. You will see that many of the questions are related to one another and that you cannot discuss one without discussing others.

Let me stress once again that there are no "right" or "wrong" answers for many of these questions. People who disagree with one another are not always "good" or "bad." There are shades of gray but perhaps some of the gray areas will become more black or white as they are discussed openly. What is needed now is for each of us to sit down and think about these and other questions so that in the future, animals' lives become better. Here are some questions to think about. Remember that there are many others.

Why do some people think it is acceptable to kill dogs while others do not?

Why do some people feel more comfortable killing ants than dogs?

Are some species more valuable or more important than others?

Do some animals feel pain, experience anxiety, and suffer while others do not?

Are some animals conscious?

Do animals feel emotions?

Are some animals smart but not others?

Do some animals have rights but not others?

What is the difference between animal rights and animal welfare?

Should endangered animals such as wolves be reintroduced to places where they originally lived?

Should we be more concerned with species and their survival than with individuals and their well-being?

Should animals be kept in captivity, in zoos, in wildlife theme parks, and in aquariums?

Should we interfere in the lives of other animals?

Should we interfere in fights in which an individual could get hurt?

Should we feed starving animals?

Should we give first-aid when they are hurt?

Should we rescue animals from oil spills?

Should we inoculate animals to protect them from diseases such as rabies?

Why do some humans eat animals?

Why do some humans use animals for research?

Why do many people feel more comfortable using dogs for research than using people for research?

Should animals be used to test cosmetics or foods?

Should domesticated animals such as dogs and cats be treated differently from their wild relatives, wolves and lions?

Do we need to cut up animals - dissect them - to learn about them or ourselves?

What types of non-animal alternatives are available for product testing, dissection, and vivisection?

Where do we go from here?

We all make choices, and the reasons why certain choices are made need to be carefully analyzed and discussed. Animals' lives must be taken seriously, and arguing that the ends justify the means - that human benefits justify the use of animals and how we treat them - is not enough.

CHAPTER 2

ANIMALS IN A HUMAN WORLD

Human and non-human primates: How close are we?

Although there are numerous differences between humans and other animals, in many important ways "we" (humans) are very much one of "them" (animals) and "they" are very much one of "us." For example, researchers have compared proteins on the surface of human and chimpanzee cells. Of nine amino acid chains (amino acids are the building blocks of proteins) studied, there are only five (0.4%) differences out of a total of 1,271 amino acid positions. This means we are 99.6 % chimpanzee and vice versa. (The word "chimpanzee" means "mock man" in a Congolese dialect.) Also, humans and chimpanzees share 98.4% of their genes. Gorillas are 2.3% different from both humans and chimpanzees, and orangutans are 3.6% different from both humans and chimpanzees.

Despite the closeness between humans and our next of kin, the Great Apes (gorillas, chimpanzees, bonobos, orangutans, siamangs, and gibbons), when humans and these and other animals' paths cross, often the animals lose. Chimpanzees are used in research that causes much pain and suffering. They are strapped to chairs and unable to move, subjected to radiation that makes them violently ill and sometimes kills them, shocked in situations from which they are not allowed to escape, and injected with infections that sicken and kill them. They are also used to study diseases, such as AIDS, which they do not contract in the wild. When they are used in these sorts of experiments, they are often housed alone in small cages and suffer severe emotional stress (depression) as well as physical trauma (great weight loss, self-mutilation). Around the world there are movements such as the Great Ape Project to make their lives better and to eliminate their use altogether. Recently, the British government decided that Great Apes would no longer be used in research in Britain. Similar legislation passed in New Zealand as well. France and other European countries have signed The Treaty of Amsterdam which recognizes animals as sentient beings capable of feeling fear and pain and of enjoying themselves.

Retirement homes

What happens to the animals when researchers are finished with their research? Chimpanzees and other animals are often disposed of when no longer useful. They are "sacrificed" or "euthanized," which means they are killed, or they languish in cages year in and year out. But now there are many people who do not want these animals' lives to end or be wasted because they are no longer useful to humans. They want to build retirement homes for chimpanzees and other animals whose research careers are over, so that these animals have the best lives possible and die a natural death. They want to rehabilitate these individuals whose lives have been damaged psychologically and physically.

One of the most visible cases concerning the fate of research animals centers on the chimpanzees who were used in the United States' space program, the "Air Force chimpanzees." While some people want to continue to use these animals in biomedical research because they are no longer needed for space research - simply move them from one cage to another - others want their use to be stopped and have them live in comfort for the remainder of their lives. After all, they did what they were told to do, and often endured much pain and suffering during space research. After a TV show, *Dateline*, on which chimpanzee retirement homes were discussed, 98% of the people who responded to a survey favored the use of these or similiar residences.

One of the biggest businesses opposing retirement homes is The Coulston Foundation in Alamogordo, New Mexico. Concerning chimpanzees, their founder, Frederick Coulston, said: "Why let them retire? I won't retire. Most people I know even if they retire from a good job continue to do something good. You aren't going to let these chimpanzees just sit out there and suffer in a sanctuary."

According to the organization In Defense of Animals (www.idausa.org/), The Coulston Foundation is the only laboratory in the United States' history to be formally charged three times by the United States Department of Agriculture for violations of the federal Animal Welfare Act. In 1996, the laboratory settled the first set of charges, which involved citations for negligent primate deaths, by paying a $40,000 fine. Other charges that are still pending include many violations relating to negligent chimpanzee deaths, including two - deficient research oversight and inadequate veterinary care - that go to the heart of the laboratory's ability to conduct quality testing. The Coulston Foundation has also run afoul of other federal agencies. The National Institutes of Health (NIH) recently placed restrictions on the laboratory's NIH-funded testing, based largely on inadequate veterinary staffing. Recently, the death of a chimpanzee named Eason, who was involved in the testing of an artificial spinal device, resulted in yet another investigation. Eason and ten other chimpanzees were under the care of only a single inexperienced veterinarian. Coulston continues to be under federal investigation for violations of the AWA. It will be a very good day for animals when the Coulston Foundation is closed down.

Animals and the law: Buying and selling animal "property"

Many people, including legal experts, disagree whether animals are truly protected by existing laws. The legal status of animals as property, as mere resources or "things" for human use and consumption, means that it is extremely difficult for animals to get meaningful legal protection. Animals almost never win when people try to balance human and animal interests.

Just because there are laws that permit something to happen - it is legal, so we can do it - does not mean that the laws cannot be questioned and changed as a result of open discussions. Did you know that it is possible for people to privately own Great Apes and that there are few regulations that these people need to follow? This should not be possible, but loopholes in existing laws and relaxed enforcement allow it to occur. In Georgia, for example, two severely malnourished chimpanzees were found in an unlit cellar living with months of accumulated feces and newspapers. Just weeks before these chimpanzees were rescued, a veterinarian signed a permit saying that these wonderful apes were being kept in adequate housing conditions. Fortunately, since they were rescued, both chimpanzees have made strong recoveries.

Here is more to think about. Animals – and not only domesticated companions – can be ordered by mail. There is a publication called *Animal Finders' Guide* from which people can order chimpanzees, monkeys, various carnivores including red foxes and wolves, and even white rhinoceroses. Recently, a baby male chimpanzee was on sale for $30,000 (USA). In the July 1, 1998 issue of *Animal Finders' Guide,* baby camels and zebras, Japanese macaques, cotton top tamarins, cougar kittens, black bear cubs, Canadian lynx, and Bengal tigers were also for sale. One needs to ask what the lives of these animals will be like in unregulated captive environments. Thankfully, a new book has just been published called *The Animal Dealers: Evidence of Abuse of Animals in the Commercial Trade 1952-1997* that openly discusses people who treat animals horribly solely for economic gain.

Changing our views of animals from "objects" or "property" to subjects of a life is a very important move. Many people are working on this problem. The organization In Defense of Animals has a campaign devoted solely to this issue, and Gary Francione has written a book titled *Animals, Property, and the Law* that deals mainly with the problems that arise when animals are viewed as property.

Listening to animals and taking their points of view

How can we as humans begin to understand and appreciate the lives of other animals who are so different from us? They do not speak any known human spoken language. The noses of many animals are much more sensitive than ours, and they see and hear things differently from how we do. Some animals fly about, others swim, and some live on the edges of cliffs, underground in tunnels, or in dark caves for most of their lives.

We can make only educated guesses about the lives of other animals, but if we study them carefully and try hard to understand how they live, we can make extremely good assumptions about the nature of their lives. Just how good our assumptions are is borne out by the fact that we are able to predict very accurately the behavior of numerous animals as they go about their daily activities. I am sure that Jethro has a better sense of smell than I do. The neurons in his dog brain are wired differently, and the area of his brain that deals with odors is more developed than mine. Jethro often lifts his head to the wind to pick up odors of others who have been there, and he also spends a lot of time sniffing the ground, shrubs, and other dogs. His nose is like a vacuum, as are our own eyes. These sense organs take in what is out there and use this information to decide what to do next.

Bats are also quite different from us. They can hear ultrasound - sounds of very high frequency - and use these sounds to avoid objects and find prey while in flight. We cannot hear these sounds, but just because we cannot, does not mean that bats are responding to make-believe sounds. The main point to always remember is that other animals are different from us and also from one another, and that each has a unique view of the world. They have their own ways of living in their own worlds, and each and every one is interesting and important.

If other animals could talk . . .

The famous Austrian philosopher Ludwig Wittgenstein once claimed, "If a lion could talk, we would not understand him." However, lions and many other animals *do* indeed speak in their own ways, and if we try very hard we can learn much about what they are saying.

Just because most animals do not do things as we do does not mean that they are not as "good" as we are. Indeed, we cannot run as fast as cheetahs, see as well as hawks, swim like dolphins, or fly like birds. But, there are many things that we can do that make us uniquely human. We can build cars and computers, fly airplanes, and worry about paying taxes. So, rather than think that other animals are not as good as we are, that they are less-than-human, it is best to realize that they are different from us, but that is neither "good" nor "bad." *Animals are certainly not less-than-human.*

Anthropomorphism

In order to talk about the world of other animals we have to use whatever language we speak. So, in order to tell someone what your companion dog is feeling, you might also use words that you use when you talk about yourself or other humans. My companion dog, Jethro, might behave as if he is happy or sad and down in the dumps. If you asked me how he is feeling, I might tell you he is happy or sad. Even if you don't see him, I am sure you would have an accurate picture of what he was doing that led me to tell you about how he felt.

By saying that Jethro is happy, sad, angry, upset, or perhaps in love, I am being *anthropomorphic*. I am using human terms to describe animals' emotions and feelings. Of course, I cannot be absolutely certain that Jethro is happy, sad, angry, upset, or in love, but I have no other way to describe how what he is doing might indicate what he is feeling. In fact, I cannot really know for sure that other people are feeling what they say they are feeling, but this would be an incredibly chaotic and disorderly world if we did not rely on at least some common sense and trust!

Anthropomorphism can also be useful for getting closer to and embracing the animals we study. Of course, it is also essential to try as hard as we can to take the animals' point of view and to try to discover answers to the fascinating question of how animals interact in their own worlds and why they do so. Imagine what their worlds are like to them. What it is like to be a bat, flying around, resting upside down, and having very sensitive hearing. Or what it is like to be a dog with a very sensitive nose and ears. Imagine what it is it like to a free-running gazelle, wolf, coyote, or deer. Take advantage of what animals selflessly and generously offer to us. Their worlds are truly awe-inspiring.

I should stress that we need to be asking "what it would be like to be a particular individual from her own perspective," not merely from our *anthropocentric* - human-centered view of things. Being anthropomorphic does not ignore the animals' perspectives. Rather, using human terms to describe animal behavior allows us to understand better the behavior, thoughts, and feelings of the animals with whom we are sharing a particular experience.

CHAPTER 3

HUMANS AND A DOG IN A LIFEBOAT: WHO SHOULD BE SPARED AND WHO SHOULD BE KILLED?

Speciesism: Using group rather than individual characteristics to make decisions about who should live and who should die

People who use species membership to make decisions about how animals can be used by humans are called *speciesists*, a term coined by the British psychologist Richard Ryder. Non-speciesists do not make that distinction - they do not use species membership to make decisions about how animals can be treated. Instead, they use *individual features* of an animal.

In practice, when deciding about the types of treatment to which animals can be subjected, speciesism often is narrowly used to mean "primatocentrism" - favoring human and non-human primates - or "humanism" - favoring humans. Human superiority is often used in speciesist arguments. Humans are thought to be above and apart from all other animals. They are thought to be "higher" or "worth more" than other animals. Some people believe that only humans and other primates experience pain. However, people taking the non-speciesist perspective realize that individuals in *many* other species experience pain, anxiety, and suffering, physically and psychologically, even if these are not the same sorts of pain, anxiety, and suffering that are experienced by humans, or even other animals, including members of the same species.

Human animals use numerous other animals in many ways: for food, research, education, entertainment, and testing cosmetics and other products. Animals such as dogs and dolphins have been used in warfare. Often animals are moved from one area to another or killed because humans want to expand their own horizons - build buildings, shopping malls, or roads. Prairie dogs are routinely slaughtered to allow human sprawl. Human benefits are used to argue that the ends justify the means, that it is all right to move or to kill animals so that humans can have more space or more roads on which to drive.

There are many issues that I can discuss concerning speciesism. The best way to deal with some of the most important ones is to present what are called "thought experiments" and to let you provide answers to the questions that are raised.

Imagine the following situation:

A small boat contains four humans and a dog. The boat is far from shore and will sink if all five individuals remain in it. One individual has to be thrown overboard and face certain death. You are asked to decide which individual it will be. One easy answer - all other things being equal, which they rarely are - would be to throw the dog overboard because "he's only a dog." He will not lose as much because his life is not as full of rich experiences as those of the humans. Furthermore, the dog will not suffer as much because he will not know that he is going to be thrown over and cannot anticipate that he will drown after being in the water for a few minutes. At least that is what some people believe.

But wait, the dog will die and this loss of life is sad and regrettable. And would not the dog's guardian also grieve and suffer the loss of her dog? What if there were four humans - three young individuals and an eighty year-old man - and a young dog who had his whole life before him? Would these facts change your mind? Would you choose to throw the old man over because he has already lived a fairly complete life and has less to lose than the younger humans and the dog? What if the dog was your companion or one of the people was your friend? What if there was a senile adult and an infant with brain damage among the humans? If you think that it is wrong to intentionally kill another human being, would you choose to toss a murderer overboard if one was on the boat? What would it mean if all individuals have an equal right not to be harmed?

While this scenario is imaginary, it offers much to think about concerning how decisions are made by humans about human and animal lives. Indeed one of the Roots & Shoots groups that I lead decided that no one should be thrown overboard and that there had to be a solution that would allow *all* individuals to live. People often use species membership to decide which animals can be used for various purposes. As I mentioned before, using species membership for such decisions rather than an individual's own unique characteristics is called *speciesism*. For example, all and only humans might constitute a protected group regardless of an individual's unique characteristics. The British government has declared a ban on the use of Great Apes in research. This decision is speciesist. It has been argued that these primates deserve special treatment because of their ability to think, or their cognitive capacities. Cognitive abilities include the capacities for self-consciousness, the ability to engage in purposeful behavior, to communicate using a language, to make moral judgments, and to reason (rationality). Some philosophers even think that animals who show these capacities should be called "persons."

Speciesists often use biological closeness, or behavioral similarities to humans - similar appearance or the possession of various cognitive capacities displayed by normal (not mentally compromised) adult humans - to draw the line that separates humans from other animals. Using these criteria, most animals would not qualify for protection. But there also are some humans (young infants and mentally impaired persons) who would not qualify either, and this can be a problem for speciesists who rely on cognitive capacities. Would these humans then be considered "non-persons?"

Because of individual differences within a species, this view from the "top," a human-centered "them" versus "us" perspective, can be difficult to apply consistently. The potential problem of allowing some non-human animals to be called "persons" while not allowing some humans to be called "persons" is one with which most humans do not feel particularly comfortable. Instances of when it is difficult to decide whether an animal is a person or a human is not a person are called "marginal cases." There is much interest in marginal cases because there is much at stake especially when a human being is considered to be a non-person and could then be subjected to the same treatment as non-human animals who also are considered to be non-persons. For example, if being able to make plans for the future is considered a necessary condition of personhood, humans who are incapable of making future plans would be considered non-persons.

Evolutionary continuity

Speciesism also ignores evolutionary continuity, the view that life falls on a continuum. Charles Darwin, the famous English biologist who wrote *On the Origin of Species* in 1859, stressed evolutionary continuity in mental abilities between many animals. Darwin argued that differences in mental abilities were differences in *degree* along a continuum, and not differences in *kind*. Here is an example that might help you see what Darwin meant. Rolls-Royces and less expensive Fords are both cars. The differences between Rolls-Royces and Fords are differences in degree - they are both cars - and not differences in kind. However, Rolls-Royces and motorcycles are different *kinds* of motor vehicles. Of course, animals are *not* objects, but we can also talk about differences in degree and differences in kind. Thus, the differences in mental abilities between, for example, wolves and chimpanzees are differences in degree rather than differences in kind. This simply means that there are many similarities in the mental abilities of wolves and chimpanzees.

Language and tools

For many years, people decided that it was the use of language that separated humans from other animals. But when it was discovered that there are other animals who use language to communicate with one another, language was no longer a reliable behavior to separate humans from other animals. Of course, other animals do not use human languages to communicate, but many animals use their own complex language to tell others what kinds of food are around, where they are traveling, how they are feeling, or what they need.

There was also a time when humans thought that only humans made and used tools, so this ability was used to separate humans from other animals. But when Jane Goodall, studying chimpanzees at the Gombe Stream in Tanzania, observed tool manufacture and use in David Graybeard and other chimpanzees, tool use was no longer considered unique to humans. Now it is known that many animals use tools.

Many people are always trying to separate us from other animals. They discover activities in which humans engage but in which no other animals are known to engage, and then use these activities to claim that humans are not only special - smarter than other animals - but also unique. However, animals do things that we cannot do. Are dogs who can sniff other dogs from a distance, or bats who can use high-pitched sounds to find prey, special to the point that they are better or worth more than humans who cannot perform these behaviors?

The main point is that all animals have to adapt to being who they are and where and how they live. Each may have special skills that others lack, but none is better or worse, above or below another species. There are no animals who can program computers or practice law. But there are no humans who can fly like birds, swim like fish, run as fast as cheetahs, or carry as much weight - relative to their own body weight - as ants.

So, are humans unique? Yes, but so are all other animals. The important point that needs much discussion focuses on the question *"what differences make a difference?"* What differences among individuals mean that it is all right to use or exploit one animal rather than another? If all life is respected, then it is hard to draw the line between those individuals who can be used, harmed, or killed, and those who cannot. But from the practical point of view, especially if you agree that *sometimes* it is all right to use animals, it is important to realize that sometimes you have to make very difficult decisions.

Animal intelligence

Often people argue that humans are smarter than other animals, and differences in intelligence make it all right to use and exploit animals rather than humans. However, there is much research being conducted on animal intelligence and the results are extremely interesting. Often, eye-opening surprises emerge. Frequently, differences in degree are found when differences in kind were expected before the research was conducted.

Researchers who study animal intelligence want to know if animals are smart. Do they think or make plans for the future? Do they try to deceive one another? Do they display signs of culture? Researchers who study these questions, especially in the wild, call themselves "cognitive ethologists."

"Being smart" means showing flexible behavior in new and unpredictable situations, and anticipating and planning for the future. When an environment changes and animals need to adjust and fine-tune their behavior to new situations, or when a novel solution is needed, thinking and planning are probably used in many instances.

What are some specific behavior patterns that indicate an individual is smart? Animals are said to be smart if they perform such tasks as counting objects, forming concepts in which differences or similarities are recognized, avoiding cagey predators, locating hidden food, making and using tools, deceiving others, or using complex forms of communication. Many diverse animals display several of these and other skills.

Studying communication is a useful way to learn about animal intelligence and the workings of their active minds. Prairie dogs and vervet monkeys use different alarm calls for different predators. Dr. Con Slobodchikoff, at Northern Arizona University, found that Gunnison's prairie dogs have different alarm calls for hawks, humans, coyotes, and domestic dogs. Likewise, vervet monkeys have different calls for snakes (pythons) who hunt on the ground, carnivorous mammals such as leopards, and martial eagles who hunt while flying. Dorothy Cheney and Robert Seyfarth, at the University of Pennsylvania, discovered that different warning calls are used to alert other group members to specific types of immediate danger. Other group members respond appropriately (even when predators are unseen), fleeing into trees and climbing out onto small branches when they hear a leopard alarm call, moving into thick vegetation when an eagle alarm call is heard, and standing on their hind legs and looking around when a snake call is heard.

Alex, an African grey parrot who has been studied by Irene Pepperberg at the University of Arizona, understands the concepts of "same" and "different" and can answer questions about the number of objects present, their color, shape, and composition. When presented with plastic objects, three yellow, one purple, and one red; and one piece of green wood and asked "What material is green, Alex?", he answers, "wood." And when asked, "How many yellow?" Alex says "three." Certainly, research on Alex would attest to the fact that calling someone a bird brain is not necessarily an insult.

There is also a lot of exciting research being done on "ape language." Roger and Debbi Fouts, at Central Washington University, have studied a chimpanzee named Washoe for years (Washoe means "person" in a Native American dialect). Washoe has mastered American Sign Language (ASL) as have other chimpanzees, and can communicate well with the Fouts' and also with other chimpanzees. The chimpanzees even "talk" to one another when away from humans and during play.

Kanzi, a male bonobo (pygmy chimpanzee) has learned to communicate using a lexigram, or keyboard, on which there are numerous symbols. Because he cannot produce words (talk), when he wants something, Kanzi points to a symbol. When he is asked a question, he answers by pointing to a symbol. Kanzi can answer questions about novel situations and comprehend sentences, even nonsense sentences, spoken to him by his mentor, Sue Savage-Rumbaugh, at Georgia State University. For example, Kanzi can understand "take the vacuum outside" or "put the ball in the microwave."

So what? What do animals do in the wild? Is bird and ape intelligence seen only in well cared for and pampered captive animals who live with humans? Not at all. Groups of Kanzi's relatives, wild bonobos living in dense forests, have been observed to disband into small groups during the day and reunite at night. They can track and locate one another by using signs that serve as symbols for where they are heading. When trails cross, the leaders stamp down vegetation or place large leaves on the ground pointing in the direction of travel. Trail notes are left only where trails split or cross, where there could be some confusion about direction, and not at arbitrary points. When all members travel together, trail markings are not used.

When Savage-Rumbaugh used trail signs that bonobos had left, she found her way to the group at the end of the day! Clearly, bonobos are communicating with one another using trail symbols. Are they mind-reading? Well, they seem to anticipate what other bonobos will do after seeing and processing the information contained in the symbolic trail markers.

There are many examples of animal deception, especially about food. Deception may include remaining silent or hiding an object when others are around, distracting others' attention by looking away from an object, or leading others away from an object. When rhesus monkeys find food, they will often withhold this information and not call others to the area. Wolves often deposit and retrieve stored food only when there is no audience. Why share if you do not have to?

Speciesism and animal intelligence

While there are species differences in behavior, behavioral differences in and of themselves may mean little for arguments about animal protection. Indeed, it is to be expected that species differences in behavior will be the rule rather than the exception, but these variations should not be viewed as being "good" or "bad" or used to place animals "higher" or "lower" on a scale of life.

Now, what about speciesism and animal intelligence? Are chimpanzees "smarter" than mice or dogs, for example? If we believe that this is so, then we need to be very clear about why.

We need to be clear about what we mean when we claim that the social lives of chimpanzees are more complex than those of mice or dogs, or that chimpanzees are able to solve more complex or difficult problems, or that chimpanzees show more flexible patterns of behavior in response to environmental changes - they can change their behavior so that they are able to survive in many different conditions. It is also important to remember that mice and chimpanzees do well in their own worlds, and neither would do well in the other's. Although some people would probably not have much trouble deciding to harm or to kill a mouse rather than a chimpanzee if they *had* to make a choice, their decision should not be made conveniently along species lines.

"Smart" and "intelligent" are words that are often misused: dogs do what they need to do to be dogs - they are dog-smart in their own ways. And monkeys do what they need to do to be monkeys - they are monkey-smart in their own ways. Neither is necessarily smarter than the other. The misunderstanding and misapplication of the notions of smartness and intelligence can have significant and serious consequences for animals, especially if they are thought to be dumb and insensitive to pain and suffering.

Drawing moral boundaries at the species level using behavior patterns that are typical of a species is very difficult. Indeed, in some instances, for example when considering whether to use a healthy mouse rather than a severely mentally impaired chimpanzee in a persistent vegetative state in an experiment, or debating restraining the physical movements of a healthy mouse rather than a severely physically disabled chimpanzee, it might be argued that an individual mouse rather than an individual chimpanzee should be spared. In decisions such as these it is not necessary that a *normal* mouse has to be compared to an *abnormal* chimpanzee. For example, if some procedure could be carried out harmlessly on a chimpanzee but would require harming a mouse, then the mouse should be spared. Thought experiments such as these can help people come to terms with the difficult issues at hand on a case-by-case basis rather than merely saying that chimpanzees should *always* be spared rather than mice.

Much animal use is driven by the similarities rather than the differences between humans (us) and other animals (them). If "them" who are used are so much like "us," much more work needs to be done to justify certain practices. Why do some people feel comfortable subjecting animals to experimental research that will harm or kill them, but refuse to use humans for this type of work? If the animals are vastly different from us, then the results will be difficult to apply to humans. However, humans are more similar to humans than they are to any other animals, and it truly is a compromise to use other animals rather than humans for research that will solely benefit people.

Why are there differences in attitudes among people? This is a very difficult question to answer, but there are a few reasons that are usually given. First, some people think that humans are special animals who are superior to all other animals because only humans have been created in the image of God and only humans are rational beings; that is, they can reason. While it is correct to say that humans are unique, it is also correct to say that all individual animals are in some ways unique, even identical twins. Other people simply feel that animals are here for humans to dominate and to control, while some believe that humans can use animals for whatever purpose they want as long as there are human benefits. For these people, the ends - human benefits - justify the means and the costs - the use and abuse of animals. A belief that runs throughout all of these views is that humans are "higher" or "better" than other animals. Although there are numerous differences between humans and other animals, it is clear that in many important ways, "we" are one of "them" and "they" are one of "us."

The Great Ape Project: Granting apes legal rights

In 1993, a book titled *The Great Ape Project: Equality Beyond Humanity* was published. This important project launched what has become known as the Great Ape Project (GAP). I was a contributor to the GAP, and strongly supported its major goal, namely that of admitting Great Apes to the *Community of Equals* in which the following basic moral rights, enforceable by law, are granted: (1) the right to life, (2) the protection of individual liberty, and (3) the prohibition of torture. In the GAP, "equals" does not mean any specific actual likeness, but equal moral consideration.

Some people do not think that the Great Ape Project goes far enough because of its speciesist concentration on Great Apes to the exclusion of other animals. I agree. But the Great Ape Project had to start somewhere, and beginning with animals that would generate the least resistance was probably the correct place to begin. While many people might be willing to legally grant certain rights to Great Apes, many would not want to legally grant rights to dogs, cats, birds, mice or other rodents (many of whom are used in research), fish, crocodiles, lobsters, or ants.

Going beyond apes:
The Great Ape /Animal Project

When I first contributed to the GAP, I wanted to include all other animals in this project and to expand the GAP to The Great Ape/Animal Project, or the GA/AP, and to expand membership in the Community of Equals. In the GA/AP, it is assumed that all individual animals have the right to be included in the Community of Equals. *All life is valuable and all life should be revered.*

It is important to include more species in discussions of animal protection and animal rights. As we learn and understand more about the animals with whom we share this planet, we will come to appreciate and respect all life even more than we do now. We need to study other animals' behavior to learn about how they live in their own worlds and about their capacities for pain and suffering.

DO OTHER ANIMALS EXPERIENCE PAIN, ANXIETY, AND SUFFERING?

their own worlds and about their capacities for pain and suffering.

One of the most basic questions with which many people are concerned deals with animal pain. Pain is an unpleasant sensation or range of unpleasant sensations that can protect animals from physical damage or threats of damage. For example, when animals are bitten hard, they move away from the animal biting them. To experience pain an individual must have at least a simple nervous system. There is no doubt that many animals experience pain. Veterinarians have developed a pain patch for dogs coming out of surgery. If they did not know that the dogs felt pain and suffered, why would they have developed this patch? Indeed, we have all heard dogs yelp when they step on nails, catch their tails in doors, or are bitten too hard. As a significant step in the right direction, The University of Tennessee College of Veterinary Medicine recently established the Center for the Management of Animal Pain to improve methods of preventing and treating pain in animals.

While there are obvious differences in the behavior of individuals belonging to different species, there are also differences in the behavior of members of the same species. However, in and of themselves, behavioral patterns may mean little for arguments about animal protection. Many animals experience pain, anxiety, and suffering, physically and psychologically, when they are held in captivity or subjected to starvation, social isolation, physical restraint, or presented with painful situations from which they cannot escape. Even if it is not the same sort of pain, anxiety, or suffering that is experienced by humans, or even other animals, including members of the same species, an individual's pain, suffering, and anxiety count.

In everyday life, the experience of pain is unavoidable. Pain serves many useful functions and contributes to survival. Humans

and animals lacking pain systems by accident of birth or di
tend to have shorter lives. While researchers are not sure w.....
animals feel pain, there is much evidence that animals who many
people thought could not feel pain, such as fish, do feel pain.

Fish, for example, have nerves similar to those that are associated
with the perception of pain in other animals. Fish show responses to
painful stimuli that resemble those of other animals, including humans
(www.enviroweb.org/pisces). Even some invertebrates, animals
without backbones such as insects, seem to experience pain and also
possess nerve cells that are associated with the feeling of pain in
vertebrates, animals such as humans, who have backbones. Whether
some insects actually feel pain is not known, but because they might,
some people believe that they should be given the benefit of the doubt.
Humans should assume that animals can experience pain, and treat
them accordingly. Even octopuses, with their large central nervous
system and complex behavior, have been given the benefit of the
doubt in Great Britain. As of 1993, they are protected under the
Animals (Scientific Procedures) Act of 1986 that regulates the use
of animals in scientific research.

While we all try to avoid painful situations, as do many other
animals, some pain may be unavoidable or beneficial for an individual, such as the pain experienced from receiving an injection that
helps to cure a disease or to prevent rabies.

René Descartes (1596-1650), a French philosopher, believed that
animals were robots and that they did not think or feel pain. Of
course he was wrong. But, some people still believe that animals
other than humans, even our close relatives such as gorillas and chimpanzees, do not experience pain. When some people are willing to
admit that other animals feel pain, it is usually those animals who
look or behave like us who are most likely to be granted the capacity
to feel pain. Companion animals with whom we are familiar, including canaries, parakeets, dogs, cats, and perhaps urban wildlife such as
deer and raccoons, also might be thought of as capable of feeling pain.

Animal pain and human pain

A major problem in knowing which animals might feel pain and
which might not is that many animals may not experience pain similar
to the ways in which humans do. We do not know if lobsters (who
some people casually drop into pots of boiling water) and other
animals we habitually abuse feel pain in their own ways, even if they

do not experience or express their pain as we do. Different animals might have more tolerance for some situations and less for others. Because they are not like us, some people think animals do not feel pain because their pain is not like ours. However, there are differences among species, and it is wrong to assume that dogs, cats, birds, fish, ants, and lobsters will behave like humans do in painful situations. Remember, we do not do many things in the ways that other animals do, and there is no reason to think that we will all feel pain or respond to it in the same way. Different species differ in many ways, including how they perceive and feel pain and how they react to it. We should not use ourselves as the measure for other species.

CHAPTER 5

ARE ANIMALS CONSCIOUS?

Many people want to know if animals are conscious: are they awake and aware of their surroundings, are they perceptually conscious? If "being conscious" means only that one is aware of his/her surroundings, then many animals are obviously conscious. Of course, numerous animals, including humans, act like robots in many situations, so it would be wrong to think that humans are *always* acting consciously.

There are different degrees of consciousness. In addition to perceptual consciousness, there is also what some call a higher degree of consciousness, namely self-consciousness, an awareness of who you are in the world. For example, as long as my brain works normally, I know that I am Marc Bekoff, and I can be fairly sure that there are no other Marc Bekoff's who are exactly like me, with the same set of past experiences. If something happens to me that I like or dislike, I know it is happening to me.

It is also possible that I may not know who I am, but I may be fully aware of something happening to my body. If I receive a blow to my head I may not know my name, or who I am, but I am able to feel the pain that is caused by my injury, and it would be wrong to make me suffer just because I do not know my name. There is little doubt that many other animals know when something painful happens to their bodies. Even if they do not know their names, they are able to experience "something bad is happening to this body."

While it is not known if other animals know *who* they are, some animals, especially chimpanzees and various monkeys, have been shown to use their mirror image to groom parts of their bodies that they cannot see without the mirror (for example, their teeth and backs). Some also look into a mirror and touch a spot that was placed on their foreheads when they were sedated, and unaware that the spot was placed there. This self-directed behavior suggests that they might have a sense of their own bodies, that this is "me." Right now we do not know which animals are aware of themselves and which are not. While it may seem obvious to some humans that animals (for example, gorillas or dogs) who are closely related to us or with whom we are familiar are self-conscious, we really do not know for sure that this is so.

If we pay attention to some basic and well-accepted biological ideas, though, it is impossible to justify the belief that we are the only species on this planet in which individuals are self-conscious. As I mentioned before, Charles Darwin showed that there are many connections among different animals, that there is *continuity* in evolution. Even if we are very different from dogs or cats, there is no reason to think that dogs, cats, and many other animals do not think in their own ways, and that they are not conscious and do not feel pain and suffer in their own ways.

Sentience: Consciously feeling pleasure and pain

Animals who are conscious and aware of pleasure and pain are said to be *sentient*. As I have emphasized before, when we are concerned with other animals, we must try to understand their own worlds - what it is like to be a gorilla, a dog, a bat, a robin, or an ant. *Their own worlds are no less important than our worlds, just as their own pain and suffering are as important to acknowledge.* Sentience is very important to consider when we make decisions about how we interact with animals.

Animal emotions

It is hard to watch elephants' remarkable behavior during a family or bond group greeting ceremony, the birth of a new family member, a playful interaction, the mating of a relative, the rescue of a family member, or the arrival of a musth male, and not imagine that they feel very strong emotions which could be best described by words such as joy, happiness, love, feelings of friendship, exuberance, amusement, pleasure, compassion, relief, and respect. (Joyce Poole. "An exploration of a communality between ourselves and ele-phants")

A greylag goose that has lost its partner shows all the symptoms that John Bowlby has described in young human children in his famous book Infant Grief *. . . the eyes sink deep into their sockets, and the individual has an overall drooping experience, literally*

letting the head hang . . .(Konrad Lorenz. *Here I Am - Where are You? The Behavior of the Greylag Goose*)

Elephant joy and sadness, chimpanzee and goose grief, and happiness and love in dogs - many animals seem to experience fear, joy, happiness, pleasure, shame, embarrassment, resentment, jealousy, rage, anger, love, compassion, respect, sadness, despair, and grief. They may even have senses of humor. Animals' emotional states are easily recognizable. Just look at their faces, their eyes, and the way they carry themselves. Even people with little or no experience observing other animals usually agree with one another on what an animal is most likely feeling. And their intuitions are borne out because they use their understanding of animal emotional states to predict future behavior rather accurately. The expression of various moods in animals raises many challenging questions about their emotional lives. As examples, I will consider grief and joy.

Many animals display profound grief at the loss or absence of a close friend or loved one. Jane Goodall observed Flint, a young chimpanzee, withdraw from his group, stop feeding, and die of a broken heart after his mother, Flo, died. Flint remained for several hours where Flo lay, then struggled on a little further, curled up, and never moved again. Here is Goodall's description from her book *Through a Window: My Thirty Years with the Chimpanzees of Gombe:*

Never shall I forget watching as, three days after Flo's death, Flint climbed slowly into a tall tree near the stream. He walked along one of the branches, then stopped and stood motionless, staring down at an empty nest. After about two minutes he turned away and, with the movements of an old man, climbed down, walked a few steps, then lay, wide eyes staring ahead. The nest was one which he and Flo had shared a short while before Flo died . . . in the presence of his big brother [Figan], [Flint] had seemed to shake off a little of his depression. But then he suddenly left the group and raced back to the place where Flo had died and there sank into ever deeper depression . . . Flint became increasingly lethargic, refused food and, with his immune system thus weakened, fell sick. The last time I saw him alive, he was hollow-eyed, gaunt and utterly depressed, huddled in the vegetation close to where Flo had died . . . the last short journey he made, pausing to rest every few feet, was to the very place where Flo's body had lain. There he stayed for several hours, sometimes staring

and staring into the water. He struggled on a little further, then curled up - and never moved again.

Here are some other examples. Sea lion mothers, watching their babies being eaten by killer whales, wail pitifully, anguishing their loss. Dolphins have been seen struggling to save a dead infant, and they mourn afterwards. And elephants have been observed standing guard over a stillborn baby for days quiet and moving slowly with their heads and ears hung down. Joyce Poole, who has studied wild elephants for almost two decades, notes that grief and depression in orphan elephants are real phenomena. Orphan elephants who saw their mothers being killed often wake up screaming.

Animals also experience immense joy when they play, greet friends, groom one another, are freed from confinement, and perhaps while they watch others having fun. Joy is contagious.

Animals tell us they are happy by their behavior - they are relaxed, walk loosely as if their arms and legs are attached to their bodies by rubber bands, smile, and "go with the flow." They also speak in their own tongues - purring, barking, or squealing in contentment. Dolphins chuckle when they are happy. Greeting ceremonies in African wild dogs involve cacophonies of squealing, propeller-like tail wagging, and bounding gaits. When coyotes or wolves reunite, they gallop toward one another, whining and smiling, their tails wagging wildly. Upon meeting, they lick one anothers' muzzles, roll over, and flail their legs. They are jubilant. When elephants reunite, there is a raucous celebration. They flap their ears, spin about, and emit a "greeting rumble." They are so happy to see one another.

Joy abounds in play. Animals get so immersed it is said, "they are the joy and the play." They show their delight by their acrobatic movements, gleeful vocalizations, and smiles.

There is a feeling of incredible freedom in the flow of play. Violet-green swallows soar, chase one another, and wrestle in the grass. I saw a young elk in Rocky Mountain National Park run across a snow field, jump in the air and twist his body while in flight, stop, catch his breath, and do it again and again. Buffaloes have been seen playfully running onto and sliding across ice, excitedly bellowing "Gwaaa" as they did so.

The more we study animal emotions and the more open we are to their existence, the more we learn about their fascinating emotional lives. Surely, it would be narrow-minded to think that humans are the only animals who have evolved to experience deep emotional feelings.

CHAPTER 6

ANIMAL RIGHTS AND ANIMAL WELFARE

What does it mean if animals can feel pain and experience deep emotions? If animals feel pain and are able to suffer, then we must be careful not to cause them unnecessary pain and suffering. While some people believe that it is all right to cause animals pain if the research helps humans, there are others who believe that this should not be done even if humans might benefit from the research.

People who believe that it is wrong to cause animals any pain and suffering, and that animals should not be eaten, held captive in zoos, or used in painful research, or in most or any research, are called *rightists*. They believe that animals have certain moral and legal rights that include the right not to be harmed. Those people who believe that we are allowed to cause animals pain, but that we must be careful not to cause them excessive or unnecessary pain, argue that if we consider the animals' *welfare* or *well-being*, then that is all we need to do. These people are called *welfarists*.

Many people support a position called the *rights* view. According to the lawyer and animal rights advocate, Gary Francione, to say that an animal has a "right" to have an interest protected means that the animal has a claim, or entitlement, to have that interest protected even if it would benefit us to do otherwise. Humans have an obligation to honor that claim for other voiceless animals just as they do for young children and the mentally impaired. So, if a dog has a right to be fed, you have an obligation to make sure she is fed, and you are also obligated not to do anything to interfere with feeding her.

Professor Tom Regan, a Professor of philosophy at North Carolina State University, is often called the "modern father of animal rights." His book *The Case For Animal Rights*, published in 1983, attracted much attention to this area. Advocates who believe that animals have rights stress that animals' lives are valuable in and of themselves and not because of what they can do for humans or because they look or behave like us. Animals are not property or "things," but rather living organisms, subjects of a life who are worthy of our compassion, respect, friendship, and support. Rightists expand the borders of animals to whom we grant certain rights. Thus, animals are not "lesser" or "less valuable" than humans. They are not property that can be abused or dominated. Any amount of animal pain and death is unnecessary and unacceptable.

Many people think that the *animal rights* view and the *animal welfare* view are the same. They are not, as Gary Francione points out in *Rain Without Thunder: The Ideology of the Animal Rights Movement*. Most *welfarists*, as people who support animal welfare are called, do not think that animals have rights. Some welfarists do not think that humans have rights either. Rather, they believe that while humans should not abuse or exploit animals, as long as we make the animals' lives comfortable, physically and psychologically, then we are taking care of them and respecting their welfare. Welfarists are concerned with the quality of animals' lives. But welfarists do not believe that animals' lives are valuable in and of themselves, that it is just because animals are alive that their lives are important.

Welfarists believe that if animals experience comfort, appear happy, experience some of life's pleasures, and are free from prolonged or intense pain, fear, hunger and other unpleasant states, then we are fulfilling our obligations to them. If individuals show normal growth and reproduction, and are free from disease, injury, malnutrition and other types of suffering, they are doing well.

This welfarist position also assumes that it is all right to use animals to meet human ends as long as certain safeguards are used. They believe that the use of animals in experiments and the slaughtering of animals for human consumption are all right as long as these activities are conducted in a humane way. Welfarists do not want animals to suffer from any unnecessary pain, but they sometimes disagree among themselves about what pain is necessary and what humane care really amounts to. But welfarists agree that the pain and death animals experience is sometimes justified because of the benefits that humans derive. The ends - human benefits - justify the means - the use of animals – even if they suffer, because the use is considered to be necessary for human benefits.

Do domesticated animals deserve less than their wild relatives?

One position that many scientists take favors the use of domesticated, or human-engineered animals, such as companion dogs and cats or various rodents, rather than wild animals in research. Some feel humans owe less to these animals, and because they can breed and care for them more easily than their wild brothers and sisters, they may be used. I believe that treating domestic animals less respectfully than their wild counterparts is unjustified. J. Baird Callicott, a philosopher who has written extensively about animal and environmental ethics, once claimed: "Domestic animals are creations of man. They are living artifacts." He also wrote "They have been bred to docility, tractability, stupidity, and dependency. It is literally meaningless to suggest that they be liberated." Callicott believes that domestic animals have been bred to be stupid but gives no indication of what measures he uses. He also does not seem to realize that simply because animals do things in ways that may seem stupid to us, this is no reason to demean them as they adapt to their own worlds. I would argue that there are not any stupid animals, only, perhaps, narrow-minded humans who do not take the time to learn more about the animals whom they call stupid.

Because some people believe that domestic animals are human creations and therefore are less deserving of our compassion and respect than their wild relatives, they argue that wild animals should be assigned higher moral status than domestic animals. Is this fair to domesticated animals whom we have created and many of whom have developed a deep trust in us? Domestic animals certainly can and do experience pain and suffering, and there is no evidence that their pains or sufferings are very different from, or less than, those of closely-related wild relatives.

Perhaps humans actually have special obligations to domesticated individuals. Possibly, domesticated animals actually suffer psychologically more than their wild counterparts when their expectations are not met in their interactions with humans. The philosopher L. E. Johnson, summed it up nicely when he wrote:

"Certainly it seems like a dirty double-cross to enter into a relationship of trust and affection with any creature that can enter into such a relationship, and then to be a party to its premeditated and premature destruction." Some field workers do indeed believe that the animals they study come to trust them. For example, Jane Goodall believes that her relationships with the chimpanzees she intensively studied "can best be described as one of mutual trust." Trust and expectations of certain types of behavior on the part of animals are brought forth by the ways in which humans have interacted with them in the past.

CHAPTER 7

UTILITARIANISM:
TRYING TO BALANCE THE COSTS AND
BENEFITS OF USING ANIMALS

People who consider animals' usefulness to humans are called *utilitarians* and they practice *utilitarianism*. The philosopher Peter Singer, author of *Animal Liberation*, who now teaches at Princeton University in New Jersey, is the modern-day champion of utilitarianism as it relates to how humans use other animals.

Utilitarians typically believe that neither animals nor humans have rights. There are no moral rules that guide the decisions that utilitarians make except to maximize utility or use-value. Utilitarians believe that a dog, cat, or any other animal can be used as long as the pain and suffering that the animal experiences - the *cost* of using the animal to the animal - is *less* than the benefits to humans that are gained by using the animal.

Singer believes that the best course of action is the one that has the best consequences, on balance, for the interests of all those who are affected by a particular decision to do something or not to do something. It is important to note that the interests of animals must be given equal consideration with those of humans, and that animals and humans have an interest in avoiding suffering.

When utilitarianism is applied to animals it is similar to *welfarism*. The only rule, and it is not a moral rule, is that it is all right to use animals if the relationship between the costs to the animals and the benefits to the humans is such that the costs are less than the benefits. Utilitarians may argue that it is all right to use one million mice in cancer research to save only one human life because the costs to the mice are less than the benefits to the human(s) who might use a treatment that was developed using the mice. Or they may believe that it is all right to keep gorillas or other animals in cages in zoos because the costs to the animals are less than the benefits to the humans who learn about the animals' lives.

One major problem with utilitarianism is how to calculate costs and benefits. How does one decide that the pain, suffering, and lives of one million mice cost less to the mice than the benefits that are gotten by one or more humans? Why not balance one million mice with one hundred humans, or one million mice with one hundred thousand humans? Because it is humans who are making the decisions about costs and benefits, there is always the chance that there will be some bias in favor of humans. One reason why many people are not satisfied with utilitarianism is just that - because it is humans who make the decisions, it is pretty easy to make the equation come out in favor of the humans. An animal's interest can be ignored if it benefits us to do so.

According to the standard version of utilitarianism, first offered by the English philosopher Jeremy Bentham, what really matters is pleasure or pain. Bentham was very interested in animals and wanted animals to be included in moral decisions made by humans. Because of his concern with animals, he wrote: "The question is not, Can they *reason*? nor Can they *talk*? but, Can they *suffer*?" For Bentham, it really did not matter if animals could think or if they were smart but rather whether or not they could suffer. It is the costs associated with suffering that need to be considered when deciding how costs and benefits are balanced. Utilitarians who follow Bentham's ideas judge an action as right if it leads to greater pleasure than pain. In other words, people should aim to maximize pleasure - the benefit - and minimize pain - the cost.

Utilitarianism has a lot of appeal because it is very flexible. But its flexibility may also mean that the utilitarian does things that are at odds with accepted morality. One way in which people argue against utilitarianism is to show that it can lead to such conclusions as it is all right to harm animals, break promises, tell lies, and even kill a human so that you can give her home to some worthy cause if the costs are less than the benefits - if the consequences for all involved have a positive effect that is greater than the negative effect.

Many utilitarians believe that the pain and suffering of most animals could never equal the benefits that humans gain from animal use, so animals start, and usually remain, behind humans when costs and benefits are decided. Some people think that the life of a chimpanzee is more valuable than the life of a mouse or a shrimp, so they may decide that fewer chimpanzees can be used than mice, for example, in biomedical research. But trying to place an exact value on the life of an animal is very difficult, and in the end, most people make decisions as if human life is always more important than animal life.

Utilitarianism at work

Let us return to an exercise similar to the dog-in-the-lifeboat thought experiment that we considered in our discussion of speciesism. Imagine that you need to decide if it is all right to use mice or chimpanzees in a research project on lung cancer, the results of which *might* save human lives. I say might because using animal models in studies of human disease does not always work to produce solutions to a problem. However, the results of the research might turn out to help numerous humans. Or perhaps the results might help mice and chimpanzees but not humans, or they might produce information that is beneficial to chimpanzees and humans but not to mice.

You need to decide if it is all right to use one thousand mice or one thousand chimpanzees in the project, not knowing if any humans, mice, or chimpanzees will benefit. But you do know that the mice and chimpanzees will experience pain and might even die from the disease, or have to be killed to have their organs and cells analyzed. What would you do? Here are some options, and I will let you formulate others.

You might simply decide that it really does not matter how many mice or chimpanzees are used even if the results turn out to have no benefits for humans. Or you might decide that even if there are no human benefits, it is all right to use one thousand mice but only one hundred chimpanzees because chimpanzees are smarter than mice, because chimpanzees will suffer more than mice, or because chimpanzees are smarter and will suffer more than mice. Of course, at some time you will have to defend why you chose the numbers one thousand and one hundred.

But there are other possibilities. Perhaps your best friend would decide that we need more information about whether there is any chance that the research will benefit humans, and until we know this, we should not use any mice or chimpanzees. Or perhaps she will decide that only mice but not chimpanzees can be used.

Still, there are other options. Would it make a difference to you if a parent, sister, or brother might benefit from the research? Perhaps it would, and you might decide that researchers can use any number of mice or chimpanzees or any other animal for that matter if there was any chance that your family member would benefit. Perhaps if your best friend was suffering from this disease, you would make the same decision.

As I said previously, because there are so many factors that enter into figuring out how costs and benefits are determined, and because there are no strict rules that can be used to make decisions about costs and benefits, it is the flexibility of utilitarianism that makes it so hard to apply.

Let us list some of the variables that play a role in using utilitarianism. They include: (1) who makes the decision; (2) who might benefit; (3) which animals are used; and (4) how the animals are to be used. One factor that needs to be included but often is not concerns possible benefits to the animals who are used or benefits to other members of the same or other species. Perhaps the individual mice or chimpanzees who are used will benefit from the research, if they survive the procedures and do not suffer pain and injuries from which they cannot recover. Perhaps other mice and chimpanzees will benefit from the pains, suffering, and death of mice and chimpanzees used in research. So, it is for the good of the species that some individuals suffer or are killed. Some people believe that this is important to consider in deciding costs and benefits.

CHAPTER 8

WHO COUNTS? SPECIES, INDIVIDUALS, AND THE REINTRODUCTION OF WOLVES

Our consideration of speciesm, rights, welfarism, and utilitarianism brings us to our next topic, namely, should we be concerned with individual animals rather than groups or species of animals? To discuss this question I will briefly consider a topic that is receiving much attention nowadays, the reintroduction of animals such as wolves into areas where they have lived in the past. Reintroduction is an important topic because it raises many questions concerning the interests and rights of individuals versus species; what, if anything, do we owe to endangered species; and also because it considers the question of global biodiversity - should humans strive to maintain biodiversity, or should we allow "natural" events to take their course with as little human interference as possible?

In the United States, one of the most ambitious reintroduction programs involves gray wolves who used to live in many areas, including Yellowstone National Park. Wolves in Yellowstone and other areas were hunted and killed by humans. As a species, wolves were almost exterminated in the lower forty-eight states of the United States. They are currently listed as "endangered" by the United States government. Some people think that we should correct the actions of humans in the past and reintroduce wolves to areas where they once lived. Others disagree either because they are concerned that wolves will kill livestock and other animals or because they feel that nature should take its course and that what humans do is part of the natural process.

Some questions about reintroduction programs

What are some of the events surrounding reintroduction? Frequently, wolves who are to be reintroduced into an area are taken from another region where they live, and moved to the new area. In a sense they are "wolfnapped" from where they and their ancestors have lived. Sometimes individuals suffer or die when they are trapped, during transport, or when they are held in cages awaiting release into the new area. *Is it fair to use individuals for the good of their species?* Is it fair to move individual wolves from areas where they and other wolves have thrived and place them in areas where they might not have the same quality of life? Some people say that the animals will die anyway, so it is all right to move them. Of course, just because animals, nonhuman or human, may eventually die in one way does not justify killing them sooner in another way. (I have discussed many of these issues elsewhere in an essay titled "Jinxed Lynx" (www.bouldernews.com/opinion/columnists/mark.html) that I wrote concerning the reintroduction of Canadian lynx into Colorado.)

There are other questions that need to be considered. Sometimes, individual wolves are kept in captivity for purposes of breeding. They live in cages and never get to roam freely like other free-ranging wolves. Is this all right to do for the good of the species?

These are very difficult questions. Humans are trying to make decisions for animals who have absolutely no say in their future. We are their surrogate decision-makers and we place our values on their lives. Should these individuals suffer and perhaps die for the good of their species? While most humans can make decisions about whether their own lives can be used for the good of their species, other animals cannot. Many of these decisions are made because it is humans who want to see or to know that wolves once again are roaming about in areas such as Yellowstone.

The story does not end here. What about other predators, animals who usually kill other animals for food, who might now experience increased competition for food? For example, wolves are killing coyotes in Yellowstone now that wolves have been relocated. Coyote packs are disappearing, and coyotes are being forced to leave areas where they have lived for decades. Robert Crabtree and Jennifer Sheldon, who have conducted long-term research on coyotes in the Lamar Valley in Yellowstone, report that wolf-killed coyotes during the winters of 1997 and 1998 resulted in a fifty percent reduction in coyote numbers. Coyotes' pack size also decreased from an average of six coyotes to four. Allowing wolves to kill coyotes suggests that wolves are more valuable. Do you agree that this is so? And if so, why?

It is also important to consider prey species such as elk - wolf food - who now will be eaten, when in the past, in the absence of wolves, this occurred very rarely. What about their pain and suffering?

Whose interests are more important, ours or the animals? Redecorating Nature

Clearly, there are many difficult questions that need to be considered. *One of the major questions is when is it all right to override an individual's life for the good of his or her species?* It is this and other questions that we will have to face in the future, as people attempt to recreate animal communities in areas where the animals once lived.

I raise these questions not because I am against the reintroduction of animals into areas where they once lived. Indeed, the Yellowstone project appears to be successful in that wolves are breeding and the population is growing. Rather, I ask them because the issues are not as clear as some people want them to be. When trying to restore ecosystems or redecorate nature, we should be concerned with the animals who are involved, not only our own human-centered goals. Individual animals should not suffer and die unnecessarily because of what we want. Ecosystems that have continued to develop in the absence of predators will also be changed. It may turn out that in some cases it would be wrong to try to regain what was lost. Furthermore, it might be impossible to recreate what once existed, simply because times have changed and we cannot reproduce what once was.

The costs of trying to recreate ecosystems by reintroducing animals might also be too great for the animals involved because of how they have to be treated and because it is unlikely that the programs will be successful; because the animals, especially those reared in captivity, are not prepared for life in the wild. Furthermore, there might not be enough suitable habitat in which they can thrive. Humans also might not accept reintroduced animals such as wolves or other predators because they are afraid of them or because they might kill livestock. These problems certainly have been important in wolf reintroduction programs. Indeed, the Yellowstone program is in jeopardy because people get upset when wolves behave like the natural predators they are. Wolves do kill livestock and do roam from the locations where they are released. If the animals are going to pay with their lives for living their evolved lifestyle, then reintroduction programs cannot be easily justified.

My dog, your dog, or the last wolves: Friendship and trust

Reintroduction programs also raise questions concerning how we make decisions to use animals belonging to different species. Here is an interesting scenario for you to consider: My friend and philosopher Ned Hettinger of Clemson University, during discussions about animal rights, environmental ethics, and biodiversity, often asks me the following difficult question:. If I am driving my car and have to hit and instantaneously kill either the last wolf on earth or my companion dog, Jethro, who would I choose to kill? My answer always is that I would choose to kill the last wolf.

However, I often ask myself, am I wrong to make this choice, am I too self-centered, what about the fact that there are many other dogs but there will then be no future wolves? Do I owe something special to the wolves? But there also will never be another Jethro; he is a special friend, he is near and dear to me, and he means more to me than any other animal, including the last wolf (who if not pregnant, will never produce more wolves anyway). This friendship makes it impossible for me to be impartial.

There are many very difficult questions that require further discussion: what if the last wolf were, indeed, pregnant, what if there was a group of wolves? My answer remains that I would still choose to save Jethro's life. While I am really not sure what I would do if Jethro were not involved, but rather another dog, if the dog was one I knew, and I also knew that she had a caring human companion, then I might very well spare the dog.

So, should a dog, even your dog, be killed to spare wild animals? Do we owe our domesticated companions less than we owe wild animals? Would your answer be different if it were another dog or if the wild animals were rare and endangered? If you choose to kill your dog for the sake of saving wild animals, perhaps those who are endangered, why is maintaining biodiversity so important as to justify this decision?

I will return to some other questions about reintroduction programs in the next section. It is important to stress here that reintroduction programs are not necessarily bad ideas and always ill-fated. Rather, there are questions that need to be given serious consideration, questions that often get buried by the appeal of reintroducing beautiful animals back into areas where they once lived, and the emotions that are evoked in doing so. It is indeed wonderful to see wolves and other animals in the wild, and it feels great simply to know that they are out there. But at what cost?

CHAPTER 9

ZOOS, "WILDLIFE" THEME PARKS, AND AQUARIUMS: SHOULD HUMANS HOLD OTHER ANIMALS CAPTIVE?

Zoos have existed for a long time. Ancient Egyptians are known to have kept collections of animals. The first modern zoos existed in Europe and opened in the late 1700's. In the United States, the first European-style zoo opened in 1874 in Philadelphia. It was modeled after the London Zoo. The first public aquarium in the United States was opened in 1856 by P. T. Barnum, of circus fame. Early zoos were primarily for humans and not the imprisoned animals. They were essentially living museums.

In the United States, the American Zoo and Aquarium Association (AZA), incorporated in 1972, inspects zoos, "wildlife" theme parks, and aquariums (collectively referred to as zoos). (I put the word "wildlife" in quotations because it is fair to ask if animals in cages truly are wild in the sense that their relatives in nature are wild.) If these institutions meet AZA standards, they are approved and accredited by the AZA. There are only 183 accredited zoos, theme parks, and aquariums in the United States and about two thousand licensed zoos that are not accredited by the AZA. These unaccredited zoos, theme parks, and aquariums are legal, but the AZA has little say in how they operate and keep animals.

It is important to stress that accredited and unaccredited zoos vary greatly in quality. Some zoo experts feel that many zoo exhibits are antiquated and that only about one third could generously be called "enriched" or "naturalistic." One zoo director says that he would like to be able to change ninety-five percent of the exhibits he has visited. While there are some zoos that are trying as hard as they can to make the lives of their residents the best they can be, there are many zoos that are of very poor quality. When making decisions about whether or not zoos should exist, we must take into account the fact that not only do all captive animals have compromised lives and are kept without their permission, but also that there are some zoos where the animals are treated horribly.

The problem of "surplus" animals

In addition to the fact that captive animals lead unnatural lives that often are tragically impoverished, as pointed out in an excellent four-part series on zoos in the San Jose (California) *Mercury News* (February 7-10, 1999), some zoos sell, trade, donate, or loan unwanted or "surplus" animals to animal dealers, auctions, hunting ranches, unidentified individuals, unaccredited zoos, and game farms whose owners actively deal in the animal marketplace. The fact is that most zoo animals are like museum specimens in that they will never be freed from captivity. The animals are treated as mere property, their fate dependent on their dollar-value.

From 1992 to the middle of 1998, about one thousand exotic animals were sold as live merchandise. Much information on animal trafficking is summarized in Alan Green's book, *Animal Underworld: Inside America's Market for Rare and Exotic Species.* Green reports that a vast underground economy exists in the trafficking of rare and endangered species in the United States. Other animals are also sold or traded regularly. There are about 250 tigers in AZA-approved zoos but about six or seven *thousand* "pet" tigers confined in horrible conditions. Bears who have lived in petting zoos often find themselves shipped off to market as food after they are no longer useful. Green's stories and pictures will bring you to tears, and we should be grateful that he has brought these issues to the forefront for public scrutiny.

Naturalistic exhibits and environmental enrichment: Trying to make captive lives better

Nowadays, many zoos have what they call naturalistic exhibits, and wildlife parks provide animals with areas in which to roam that resemble their natural habitats. There is an attempt to meet the physical and behavioral needs of the caged animals by providing them with an environment that resembles as closely as possible their natural environment, one from which they may have been taken or in which wild relatives still live. However, for example, while American brown bears in captivity spend about the same percentage of time as wild bears being active, the bulk of captive bears' activity is spent pacing, not foraging. There is also a move to have animals live in naturalistic groups. Wolves often are kept in groups that resemble wild packs, and animals who are more solitary are allowed to live alone and have places into which to escape or hide if they want to get away from other animals or from human spectators.

Some zoos attempt to provide captive animals with enriched environments that stimulate and challenge the animals and reduce the boredom that results from being in the same place with little to do day in and day out. Enrichment programs give animals control over their environment and often provide them with choices of activities in which to engage. Nonetheless, all individuals in zoos have their freedom of movement constrained, and even in the most enriched environments, there are limits on the behavioral choices that can be made. Much enrichment is done for the visitors' sake, and enriched environments typically do not resemble the natural environments of the caged animals.

Many types of enrichment have been used. Different species have different needs, and there are individual differences within species (for example, age and gender may influence what works). The main goal of enrichment programs is to give animals more control of their lives and have them work to occupy their time. Enriching animals' lives can be accomplished by providing safe and secure places for resting, sleeping, and escaping from unwanted intrusions by cage mates and humans; allowing individuals of social species to live in pairs or larger groups that resemble natural groups; making them work for meals by providing frozen food or by scattering or hiding it; providing natural substrates; spraying various odors in cages; allowing or making it easy to exercise; providing large cages (although this is not always enriching); and increasing the complexity or diversity of their social and/or physical environments. Moderate levels of stress that tax the animals might be beneficial, especially for individuals who are to be reintroduced to the wild. For example, it might be helpful to make individuals work hard for food if food is going to be difficult to obtain in the wild.

While enrichment programs do not directly address the question of whether or not humans should hold other animals captive, the very few captive individuals who have the opportunity to experience enriched environments do seem to be more content or happier than those who do not. Animals who are in good psychological health - happy and content - do not engage as much as unhappy or discontent individuals in repeated and stereotyped pacing, cage-rage and self-mutilation, rocking back and forth; nor do they exhibit unusually high levels of fear and aggression. They do, however, engage in various types of play, have good appetites, and do not suffer from the abnormally high levels of stress, anxiety, or disease as their less fortunate kin.

Nonetheless, captive animals whose lives are enriched, similar to animals in cages, have their freedom of movement restricted, and their lifestyles are very limited. Individuals simply do not have the opportunity to live like their wild relatives, roaming freely and pursuing the lives that they were supposed to live. Their range of choices is very limited, and they have no control over their own destinies.

Public opinion about zoos

There is much disagreement about whether zoos should or should not exist. A 1995 Roper poll showed that almost seventy percent of Americans are concerned about the well-being of animals in zoos. Surely, existing zoos are not going to close in the immediate future, and the live animal specimens who are kept in cages are not going to be killed or released into the wild. This simply is not possible, and it would be unethical to kill the animals or to release animals who might not know how to live outside cages without human assistance. As I mentioned before, unwanted and "surplus" animals also need to be dealt with. Often they are housed in cages away from the main exhibits, or are sold or traded.

The results of a recent study are interesting to consider when discussing people's reactions to visiting zoos. It was discovered that children see similarities between humans and animals, whereas adults see differences. Children feel a kinship with many animals. The study also showed that what children might learn from going to zoos is not about the animals but about differences between themselves and adults. Furthermore, children frequently see how horrible zoos are and often express it later in life but not at the time of the visit. They express sorrow for the confined animals and see them as bored creatures.

Education, conservation, biodiversity, and endangered species

Two common reasons given to justify the existence of zoos include *education* and *conservation*. Some people believe that zoos are good because they educate people about animals in general and also about animal species that they would otherwise never get to see. However, Michael Kreger, at the Animal Welfare Information Center, found that the average visitor spends only about thirty seconds to two minutes at a typical exhibit and reads only some signs about the animals. A number of surveys have shown that the predominant reason people visit zoos is to be entertained. In one study at Edinburgh Zoo in Scotland, only four percent of zoo visitors went there to be educated, and no one specifically stated he/she went to support conservation. There is very little evidence that much is learned and retained that helps the animals in the future (see Animalearn Fact Files, Animals in Entertainment).

Some people support zoos because they might serve to keep individuals of rare, threatened, or endangered species alive when the habitat of these animals has been destroyed. However, it has been estimated that about fifty to seventy percent of orphaned gorilla infants who are taken into captivity will die. The figures are similar for orphaned gorilla infants and juveniles who are released from captivity into the wild.

Some people think that zoos are valuable because they will help maintain biodiversity. They argue that without zoos, biodiversity will decrease as species go extinct. Therefore, zoos potentially can be important in conservation efforts by keeping animals in safe places and then releasing them or their offspring into the wild. But if habitat is not preserved for them - if people use the land for other purposes while the animals are held in captivity - there will not be anywhere for them to be released. This happens quite often. Indeed, most conservation biologists agree that habitat loss is the major cause for losses of biodiversity. There are too many people and too little land for animals to thrive and survive. The situation is not getting better. Remember that in Kenya it is estimated that wild lands are disappearing at a rate of two percent a year.

Zoos actually do little to increase biodiversity. While some zoos make serious efforts in the conservation arena, few zoos actually have conservation programs, and in those that do, only a small percentage of the zoo's budget is spent on these programs. In a period of ten years, the San Diego Zoo reported that it spent $55 million on public relations but only $17.6 million on wildlife conservation studies.

There is little evidence that release/reintroduction programs using animals or their offspring who have been former residents of zoos are successful. While successful reintroductions have been performed for Arabian oryx and golden lion tamarins (in Brazil), there have not been many other programs that seem to have made any difference. Since 1900, of 145 individual reintroductions involving 126 species (while 13 million individuals were born in captivity), only sixteen (eleven percent) have succeeded. In 1995, Benjamin Beck, then Chair of the American Zoo and Aquarium Association's Reintroduction Advisory Group, lamented, " . . . we must acknowledge frankly at this point that there is not overwhelming evidence that reintroduction is successful." Beck also noted in 1996 that we just do not know enough to have successful rehabilitation and release programs of apes in captivity.

Vicki Croke, in her book *The Modern Ark*: *The Story of Zoos: Past, Present and Future,* quotes Terry Maple, director of Zoo Atlanta, as saying, "Any zoo that sits around and tells you that the strength of zoos is the SSP [Species Survival Plan] is blowing smoke." Thus, zoos' ability to save species is not a present-day reality.

As Dale Jamieson, a philosopher at Carleton College in Northfield, Minnesota observes, zoos basically are places where people can go to see animals, places to go on a weekend outing for entertainment. People can eat popcorn and watch and possibly torment animals by banging on their cages and making noises that scare them. There is really no evidence that people learn very much about animals that they remember after they leave the zoo, and there is no support for the claim that people donate much money or time to help animals after they have seen them in zoos, wildlife parks, or aquariums. It has been shown that watching wildlife videos in the comfort of home is better for learning about animals and for making people more sensitive to the plight of captive animals.

Jamieson also argues that if zoos truly wanted to use some animals for release or reintroduction, they would limit contact with people, provide them with large ranging grounds that resemble natural habitats, and prepare them to kill food in ways that viewers would find shocking. Roger Fouts believes that if zoos and wildlife theme parks truly cared about animals, they would put the well-being of animals higher on their list of priorities than public viewing. Of course, this would most likely mean that money would stop flowing in and zoos would face financial problems. Nonetheless, Fouts has an important point - if people who run zoos say the animals come first, then they should put their words into action on the animals' behalf.

Zoos as businesses: The dollar speaks

The existence of zoos and wildlife theme parks raises numerous important and difficult questions. Many ethical concerns are raised because zoos are businesses, and the bottom line centers on money. It costs an enormous amount of money to bring animals into captivity and to keep them there. It has been suggested that the money used to capture, transport, and keep animals in cages would be better used to do research to learn more about their lives in the wild and to conserve their habitat.

How animals are kept in zoos and what people get to see in zoos is influenced by available money. Large sums of money are spent on public relations and not on the animals. Susan Davis, in her well-researched book *Spectacular Nature: Corporate Culture and the Sea World Experience*, points out that Sea World's version of nature - the images of nature that are represented to the consuming public - is a manufactured corporate point of view. She wrote that "Sea World isn't so much a substitute for nature as an opinion of it, an attempt to convince a broad public that nature is going to be all right." Davis also concluded that "Sea World is a machine that profits by selling people's dreams back to them - dreams of a happy family, congenial public places free of fear, a peaceful community . . . " But most of the theme park's customers really do not believe it, except for a short-lived moment. Sea World personnel decide what is news. Announcements of new animals and new animal exhibits are much more likely to make the news than are animal rights demonstrations or deaths of animals.

Vicki Croke notes: "The zoo is not a window on nature but rather a prism that bends the light according to the culture it is set in." How money is used in zoos is closely related to a culture's view of animals.

Zoos and the domination of animals: Getting the blues and cage-rage in zoos

Zoos, wildlife parks, and aquariums, even the best of them, are for the most part examples of human exploitation and domination of animals, just as are circuses, rodeos, and most hunting and fishing. If the gates of zoos were left open, there would not be any animals in them after a very short while.

It is indisputable that many individual animals really do get the blues in zoos. They get bored, pace about, often engage in self-destructive behavior, and frequently become unhappy and depressed. They get cage-rage. As Vicki Croke concluded: "While the zoo can be an intriguing place to visit, it can be an awfully boring place to live."

Some questions to consider about zoos and the keeping of captive animals

After this brief background, what do you think about zoos and wildlife parks? Are there any benefits to the animals that justify humans keeping them in cages or in tanks of chlorinated water where their freedom is compromised? Make a list of what you think are the good and bad aspects, and think about how you would make zoos better for the animals who are there. Do you think zoos should exist, or should they be phased out? Do you believe that it matters if the animals in the zoos were born there or were captured in the wild and taken to a zoo? Do you think that animals who were once wild miss what they once had? Is it all right to keep animals in captivity even if there is a very small chance that they will ever be released to the wild? Jane Goodall has repeatedly told audiences around the world that in order to protect Great Apes, more habitat is needed. The same can be said for other animals such as wolves. The main reason there is so little suitable habitat is because there are too many people.

One problem with keeping some animals in captivity is that they need to be fed live animals. Do you think it is all right to feed other animals to meat-eating animals, individuals who have been killed or allowed to be killed so that the meat-eaters can eat? A recent study by Raymond Ings, Natalie Warren, and Robert Young of two hundred visitors to the Edinburgh Zoo in Scotland, showed all visitors said it was all right to feed live insects to lizards if it was done off-exhibit, and ninety-six percent said it was all right if on-exhibit. Seventy-two percent of the people agreed with live fish being fed to penguins on-exhibit and 84.5% if done off-exhibit. Only thirty-two percent agreed to a live rabbit being fed to a cheetah on-exhibit, whereas 62.5% agreed to this done off-exhibit. In general, females were more likely to object to the feeding of live vertebrate prey. Clearly, some of these people were thinking along the lines of there being "higher" and "lower" species - a hierarchy of species - even if they were not aware of this. Many people thought it fine to feed live prey because it was natural. If they disagreed with feeding live prey, it was because it would upset them or their children. *Ethical issues did not seem to count for much at all.*

It is essential that as long as zoos exist, we try as hard as we can to make them the best that they can be and to make the lives of all captive animals the most complete and richest possible. As animals' guardians, we owe them unconditional compassion, support, and respect.

CHAPTER 10

IS THE WILD REALLY GOOD
FOR ANIMALS?

Nature's "cruelty" and interfering in the lives of wild animals

Most people believe that only humans have obligations to other animals. Because animals do not know "right" from "wrong" or "good" from "bad," when an animal does something that humans would call "nice" - for example, caring for their own or other youngsters or interfering in a fight on behalf of a friend - we cannot say that they know they are being nice. When they do something that we would call "bad" - for example, wolves killing a moose - likewise, we cannot say that they are doing something wrong. The case of a rare white Bengal tiger who killed a zookeeper at the Miami, Florida zoo brings up some important issues concerning species differences in knowing right from wrong. After the zookeeper was killed it was decided that the tiger would not be destroyed because "the tiger was just being a tiger" (*Rocky Mountain News,* 7 June, 1994, p. 3A). He was not responsible for his actions, he did not know right from wrong. (Recall that when wolves are reintroduced to an area where they once lived, they often are killed for preying on livestock, for "being wolves.")

Philosophers refer to animals as "moral patients" and humans as "moral agents." This means that as moral patients, animals, if they have rights, are entitled to receive certain types of treatment. However, humans, as moral agents, have to act to meet these obligations, that is, those humans who are responsible for their actions. Infants and senile adults who are not responsible for their own lives cannot be held responsible for other humans' or animals' lives. They are not moral agents but moral patients.

Whenever humans interfere in the lives of wild animals, it is important to consider whether these intrusions can be justified. There are many reasons that people give for intruding into the lives of wild animals, and they range from being deeply concerned about helping wild animals to superficial and self-serving justifications that carry little or no weight. For example, the director of life sciences at Ocean Journey aquarium in Denver, Colorado attempted to justify keeping fish in tanks by claiming that "Fish have a very small world… You can think of it as they are getting three square meals a day and free health insurance." As I mentioned before, there is much research that shows that fish suffer and feel pain. Also, the claim that the world of fish is very small reflects a human-centered view of other animals' worlds. The worlds of many animals are very large, diverse, and complex. Humans' worlds are fairly small and narrow compared to those of many other animals.

So, for whose journey will this new facility provide? Certainly not the animals'. Thanks, but the animals do not need more tanks. Furthermore, if fish have such small worlds and uninteresting lives, why should anyone want to pay to go see them? With friends like this in the zoo business, animals do not need enemies.

Francis Crick, the Nobel prize-winner for his work on the structure of the genetic material, DNA, believes that "it is sentimental to idealize animals" and that life in captivity is better - longer and less brutal - for many animals than is life in the wild. The view that we are doing animals a favor by keeping them in cages or tanks is narrow-minded, anthropocentric - speciesism. If captivity is so wonderful and if confined animals are so lucky to get free meals, health insurance, and protection from nature's perils, why not incarcerate people? See how many people would volunteer. This line of reasoning can easily lead to the conclusion that if people truly want to be humane, we should place all animals in captivity or slaughter all wildlife humanely before they succumb to one of nature's horrible deaths.

Those people who argue that the wild is not the ideal state for animals because they suffer from diseases, injuries, starvation, predation, and intrusion by humans, feel that what humans provide for captive animals is better than what nature provides. But this view leans far to the side of human-centered control and management. Because animals evolved in nature, we should respect this fact. Yes, life can be tough out there, but it is all too easy to claim that because life is tough "out there," we are really doing animals a favor by keeping them "in here" and trying to enrich their impoverished lives.

It is important to keep in mind that animals did not evolve in cages, and animals in cages do not get to roam freely or make choices about how to live or face the trials and tribulations of being wild. Wild animals do experience pain and suffering, and their lives are not as glamorous as some people make them out to be. Nonetheless, they are free and are able to live the lives of free individuals.

Because of the many restrictions on the lives of captive animals, some people believe that a wolf in a cage is not truly a wolf. Of course, the animal is a wolf but she does not have the freedom to live the life of a wild wolf. That individual does not have the opportunity to live in a natural social group of family and friends or to hunt. He or she may starve, suffer from disease, or possibly be killed by a human in the wild. Of course, animals still have many of these "opportunities" in captivity. But because most stressors of living in the wild are natural, individuals are accustomed to handling them. Their behavior evolved in the wild.

While wild animals may be confronted with situations that bring them pain and suffering such as disease, predation, and aggression, except in unusual circumstances I do not think that wild animals should be interfered with. Vicki Croke claims that there is a joke in the zoo world that animal rightists would shut down the wild if they witnessed the savagery out there. I doubt this is true. Most people I know who support the rights position certainly wish that wild animals did not suffer from nature's perils, but they would not support programs that take animals out of the wild and put them into captivity to protect them from nature's ways.

The philosopher Paul Taylor has written much about this topic and developed what he calls the "rule of non-interference." According to this rule, humans have a duty "to let wild creatures live out their lives in freedom" because intrusion into "the domain of the natural world . . . terminates an organism's existence as a wild creature." While it may seem that a wolf in captivity is better off than a wolf in the wild who is starving, because of natural cycles of prey, or because she is a subordinate member of her pack, once even a starving wolf is brought into captivity she remains a wolf only because of her species membership. She is no longer a wolf in the sense of a wild being who lives the life of a typical member of her species.

We also may bring more pain and suffering to individuals' lives than they would endure in the wild by taking them into captivity, although there may be different types of pain and suffering in each location. For example, with rare exceptions, the life of a tiger is not improved by putting him in a zoo. Although his food will be provided for him, hunting has played a large role in the evolution of tigers and is essential to a tiger's way of life. His movement will also be severely restricted, and for animals who typically roam in search of food and shelter, captivity produces an impoverished existence. Furthermore, it is not at all clear that captive animals live longer or are healthier than their wild counterparts. Captive and wild apes, for example, show considerable similarities in weight, but there seems to be a greater risk for captive males to suffer from obesity when compared to wild relatives. This may be due to stress, inactivity, or boredom.

James Kirkwood, a British zoologist, presents thoughtful views on the well-being of wild animals and considers such questions as whether we should intervene on behalf of free-living wild animals; and if so, to what extent and how. While Kirkwood recognizes that there are many different views on these important matters, he claims that "Most would probably agree that when wild animals are harmed by man's very recent (in evolutionary terms) changes to the environment (such as oil-spills, power lines, roads, and environmental contamination) there is a reasonable case, on welfare grounds, to intervene." He also writes about veterinary intervention to treat injured or sick wild animals. Kirkwood calls for "an international code on intervention for wildlife welfare to provide guidance on ethics, methods, and standards."

African wild dogs: Who knows what happened?

One of the most visible and disputed examples of the possible effects of human interference into wild populations concerns the plight of African wild dogs, those white, tan, brown, and black-blotched dogs with large ears. Available information on this topic has been compiled and analyzed in a recent book titled *The African Wild Dog* edited by Rosie Woodroffe, Joshua Ginsberg, and David Macdonald. Although numerous scientists have been trying to figure out if humans were the cause of the incredible decline and decimation of some populations of these splendid animals, the situation is not clear.

Interference into the lives of wild dogs involves vaccinating them against rabies and canine distemper. While some scientists feel that handling and inoculating them is directly responsible for their decline because handling weakens the dogs' immune system, making them less resistant to stress, others, using the same data, conclude just the opposite. Here we have a useful example of bright and interested scientists, all of whom care deeply about African wild dogs, not being able to figure out what caused the decline of these animals. What should the researchers do - interfere and possibly cause animals to die, or let nature take its course? If the rabies and distemper were introduced by domestic dogs who would not have been there in the absence of people, are we more obligated to try to help the wild dogs than if the rabies and distemper were natural?

Some difficult questions concerning human interference

Examples such as these demand serious discussion. I believe that there are circumstances in which humans may have to intervene in the lives of wild animals. What do you think about this? Do you think that if any experimental manipulation performed by humans, including the mere presence of researchers, leads to harm for either the target animal or (indirectly) for any individual, we then have an obligation to intervene on the animals' behalf? Perhaps there are research projects that simply should not be done until we know how they influence the animals being studied. Do you think that this is a possible future course?

Trying to distinguish between what we should do and what nature does do is troubling, and using nature's brutality to justify the treatment of non-humans will not do. Do we really want an ethic that sanctions the treatment of animals by humans, as long as it is better than what nature typically has in store for similar individuals? I do not think so.

Another important and related question that arises frequently is whether human-caused pain in animals is less than or equal to what animals would experience in the wild, and if so, is it then permissible to inflict the pain? For many animals, it is difficult to know whether human-caused pain is less than or equal to what the animal would experience in the wild, for we do not know how most individual animals in nature experience pain. *We must be careful that using nature's supposed cruelty is not a justification for how cruelly animals are treated.*

CHAPTER II

ANIMALS AS FOOD:
THERE ARE ALTERNATIVES

One of the most common uses of animals is for food. About five million dairy cows are kept in confinement in the United States. Dairy cows are forced to have a calf every year, and are milked during seven of their nine months of pregnancy. This is extremely demanding on their bodies and on their psychological states. Their calves are removed from them immediately after birth so they do not drink their mother's milk. These dairy cows are literally milk machines, and they are not allowed to be mothers, to care for the young they have brought into the world.

Animal meat also is popular. Each year billions of animals are bred, transported, and housed in slaughterhouses waiting to be killed (and often watching other animals being brutally slaughtered). In the United States alone, 93 million pigs, 37 million cattle, 2 million calves, 6 million horses, goats, and sheep, and nearly 10 billion chickens and turkeys are slaughtered for food each year (see Animalearn Fact Files, Animal Agriculture). This translates into about three hundred animals per second being killed for the food industry. Numerous animals also die before they reach slaughterhouses. It has been estimated that more than 780 million chickens, 116 million turkeys, 1.8 million cattle, 2.8 million veal calves, 15.1 million pigs, and 1.2 million sheep die before arriving at slaughterhouses in the United States (*The Farm Report*, Spring 1997).

Poultry and eggs are now the most abundant and least expensive animal food products, because of the development of a large-scale industry devoted to poultry production. Birds are kept in tiny, barren battery cages, and cannot perform behaviors such as dust-bathing, perching, and nesting. Many birds also have their beaks seared off to reduce injuries and mortality associated with feather pecking and cannibalism. About one-half of the beak is removed, using a hot cauterizing blade or a precision trimmer. The pain associated with beak-trimming is intense and long-lasting. Caged birds often develop osteoporosis (weakened bones) because of a lack of exercise combined with calcium deficiency associated with their high rate of egg-laying. Approximately twenty-five percent of hens sustain broken bones when they are removed from their cages to be transported to a processing plant. Hens now lay upwards of three hundred eggs per year, as compared to 170 in 1925.

Food animals can suffer physical and emotional pain throughout their lives, often made worse by methods used to make them "better food." Individuals are fattened up for human consumption using various hormones and by keeping them in crowded and restricted housing conditions. Chickens used for meat consumption, or "broilers," now grow to market weight in about six rather than sixteen weeks. Many animals die from stress and disease before being slaughtered. Often fully conscious chickens and turkeys are shocked or drowned in an electrified bath of water, and scalded alive as they are being prepared for market. Pigs and cattle are supposed to be stunned before being hung upside down by their hind legs, having their throats slit, and bleeding to death, but often they are awake during the whole process.

There are also genetic engineering programs that produce animals that are bigger and more meaty. There is as little regard for the rights of these "bigger and better" animals as with their "normal" relatives. Dairy cows stimulated with Bovine Growth Hormone (rBGH) can produce as much as one hundred pounds of milk a day, about ten times more than they would normally yield. They suffer from udder infections and are treated with antibiotics as they continue to be exploited for milk. The antibiotics can be transferred to their milk and consumed by people.

Cows, grain, and human starvation

Animals raised for human consumption require a lot of food and land. For example, it takes eight or nine cattle a year to feed one average meat eater. Each cow needs one acre of green plants, corn, or soybeans a year. Thus, it takes about nine acres of plants a year for the meat one person eats, rather than half an acre if you do not eat meat. The amount of grain that is needed to feed animals in order to provide meat for one person could be used instead to feed approximately twenty people enough grain to live for a year.

In the United States alone, livestock eat enough grain and soybeans to feed over a billion people. It takes about sixteen pounds of grain to make a pound of beef. A reduction of meat consumption by only ten percent would result in about twelve million more tons of grain for human consumption. This additional grain could feed most, if not all of the humans who starve to death each year - about sixty million people!

Veal and public opinion

Veal, which comes from dairy calves who are imprisoned in cages so small that they cannot move, may be the best example of extreme animal abuse for food. Most formula-fed veal calves are raised in tiny twenty-four inch wide crates for their entire lives, sixteen to eighteen weeks, and fed a liquid diet twice a day. Prior to being killed, iron intake is restricted to below normal levels and the calves become anemic. Anemia results in a pale or white color of the meat, and it is the paleness of a carcass that is the most important factor in grading the meat and the price paid to the producer. The production and demand for formula-fed veal has dropped sharply since 1985, and has now stabilized at approximately 800,000 calves per year - a decrease of over four hundred percent. *Public outrage over how veal calves are treated is the major reason for this decline. It is very clear that what people think does matter. It is essential to keep the pressure on the veal industry.*

Hunting and fishing

The slogans "Gone Fishin' " or "Gone Huntin' " really mean "Gone killin'." According to a survey published in 1998 by Mark Duda, Steven Bissell, and Kira Young, about 14 million Americans, sixteen years and older (mainly males, but there is an increasing number of females) hunt each year; and about 36 million Americans, sixteen years and older (mainly males, but there is an increasing number of females) fish.

Much hunting and fishing is for food, but a meal is not always the goal. Some people engage in trophy-hunting, sport-hunting, trophy-fishing and sport-fishing, and kill animals for the thrill of the hunt; some stalk the animals but do not shoot at them or catch them; others catch fish and throw them back into the water. Often people who catch fish and throw them back deny that any damage is done to the fish. To satisfy the need for fish, in North America over 150 species have been introduced outside their natural range. Many fish are raised in hatcheries that are essentially zoos, where they are held captive, fed well, and protected until they are released to become prey for humans who fish. The introduction of these species has changed the ecological balance in numerous bodies of water. Introduced species have spread diseases to natural populations or have displaced or preyed on native species. Because of the detrimental effects that introduced fish have on ecosystems and their inhabitants, a new ethic urges caution when stocking is being considered.

Hunting and fishing result in killing animals using guns and other devices for which the animals have not evolved natural defenses. No animal on earth has adequate defense against a human with a gun, a bow and arrow, a trap that can maim her, a snare that can strangle her, or a human who uses a fishing lure that has been developed for the sole purpose of fooling fish. Even if fish are caught and released by people who try to justify fishing, this does not really solve any problem, because often their mouths are torn up, and they are returned to the wild weak and injured. As I have mentioned before, fish are capable of suffering in ways similar to mammals, including humans (www.enviroweb.org/pisces/). The stress responses of fish to stimuli that lead to anxiety and fear closely mimic those of other vertebrates. Furthermore, no one knows how many fish die after being caught and tossed back. It is estimated that five to ten percent of trout die from the stress of merely being handled.

Even if people only stalk animals but do not try to kill them, animals suffer during the chase. They are anxious and scared. Patrick Bateson, at the University of Cambridge in England, has shown that red deer hunted by dogs show stress responses similar to those seen when animals are frightened. Hunted animals showed high levels of cortisol from the start of the hunt, and levels were higher in deer hunted for longer distances. The deers' muscles also showed signs of physical damage, and the hunted deer showed extreme signs of fatigue. Non-hunted deer did not show the same stress responses - they had low levels of stress hormones. Clearly, the animals did not like being chased and hunted. Imagine being chased for the sake of being chased, but not knowing what will happen if and when you are caught.

Vegetarianism: A good alternative to killing animals

There are many alternatives for the vast majority of people who choose to eat meat. Vegetarian diets are much healthier than diets that contain meat, especially meat that has been injected with various types of hormones or from animals who were stressed before they were killed. Nutritionist Colin Campbell, in his long-term study of dietary habits in mainland China, has shown that a low-fat (ten to twenty percent of total calories), plant-based diet could significantly decrease the occurrence of chronic degenerative diseases such as various cancers and heart disorders that are prevalent in Western countries.

Due to the animal cruelty involved in meat eating, many people choose to reduce or eliminate their consumption of meat. They try cutting back on hamburgers and other animal products from five to two a week for a month, then from two to one for a month, then one for two months, etc. You might be doing not only yourself and animals a favor, but also be helping to feed other humans who might otherwise starve to death. Remember that a reduction of meat consumption by only ten percent would result in about twelve million more tons of grain for human consumption. This additional grain could feed most of the humans who starve to death each year - about 60 million people.

There are many types of vegetarianism and numerous reasons for becoming a vegetarian. *Lacto-ovo vegetarians* eat eggs and dairy products but no meat; *lacto-vegetarians* eat dairy products but no eggs or meat; *ovo-vegetarians* eat eggs but no dairy product or meat; *vegans* consume no meat, dairy products, or eggs; *macrobiotic vegetarians* live on whole grains, sea and land vegetables, beans, and miso; *natural hygienists* eat plant foods, combine foods in certain ways, and believe in periodic fasting; *raw foodists* eat only uncooked non-meat foods; and *fruitarians* eat fruits, nuts, seeds, and certain vegetables.

The philosopher Michael Allen Fox lists the following arguments for vegetarianism: (1) health; (2) to reduce animal suffering and death; (3) to promote impartiality and universal well-being; (4) environmental concerns; (5) to promote universal compassion and kinship with other animals; and (6) religious arguments. He notes that vegetarianism may be seen not only as a means of focusing attention on human-animal or human-nature relationships, but also on the choice of a way of life that is morally and ecologically preferable.

Some people become vegetarians because they care about human starvation. For example, a lacto-ovo vegetarian's diet would feed the world's human population more efficiently than a meat eater's diet because a cow must eat many pounds of vegetable matter to grow a pound of meat, and much of that vegetable matter could be used to feed humans.

Many people become vegetarians out of concern for the well-being of farm animals. Now, ask yourself, what can you do to make life better for animals who are used for food? If you eat meat, why do you do so? Is it out of habit or because you want it? Could you cut back on your consumption of meat of all kinds and other animal products? Can you talk to other people about the importance of respecting animals and not letting them be killed for food, especially when it is not necessary? By asking these questions, you can truly make a difference in the lives of many animals.

Great apes as bushmeat: Logging, hunting, and making music

Bushmeat, the meat of wild animals caught and killed in their home forests, is a very popular commercial food product in many parts of the world. Chimpanzee and gorilla meat is favored, as is kangaroo meat. About twenty percent of bushmeat is primate meat. Its consumption (even in elegant European restaurants) and trade is the biggest threat to biodiversity in some African forests. In the Congo Basin, bushmeat is the primary source of animal protein for the majority of families.

There simply are not enough chimpanzees or gorillas to sustain their slaughter for food. In one study, it was found that about eight hundred gorillas were killed each year in the Kika, Moloundou and Mabale triangle in Cameroon. If only three thousand gorillas live in that ten thousand square kilometer area, the taking of this many gorillas is not sustainable. About four hundred chimpanzees were also killed in this area. Thus, 1200 great apes in one small area were killed. Anthony Rose, at the Biosynergy Institute in Hermosa Beach, California, notes that this year more than three thousand gorillas and four thousand chimpanzees will be illegally butchered. That is five times the number of gorillas on Rwanda's Mt. Visoke and twenty times more chimpanzees than live near Tanzania's Gombe Stream, Jane Goodall's study area. More great apes are eaten each year than are now kept in all the zoos and laboratories world-wide.

The bushmeat trade is an example of how different human activities that seem to be unrelated actually greatly influence one another. For example, the large increase in the availability of bushmeat is a result of increased logging activities. Logging companies build new roads into areas that previously were very difficult to reach, and they allow hunters to travel in company vehicles to hard-to-reach areas where gorillas, chimpanzees, and other large animals are found. The hunters kill all but the smallest animals and carry meat to logging camps where loggers consume some of it. Remaining meat goes to market in cities. As logging increases, so does the number of individuals who are brutally slaughtered. There now are campaigns to have logging companies stop the transport of bushmeat.

The killing of great apes is illegal in every country in which it takes place, but very few hunters are ever punished. There are international laws concerning the killing of endangered species (for example, chimpanzees), but it is difficult to catch hunters in action. Although people have known about the harmful effects of the bushmeat trade for years, until recently, there has not been much interest in this activity. But now most conservationists realize that if commercial bushmeat hunting continues, there will be devastating effects on the population of chosen animals. Thus, there is a lot of interest in stopping the bushmeat trade.

Of course, logging is not only a problem in Africa. Many conservation biologists are supporting programs that limit commercial logging in numerous countries so that habitat and animals can be protected. One way to decrease the need for wood is to be careful when you buy wood products and ask questions about the source of the timber. Ask where the wood came from. If you are unsure about its origins, find out more. It is possible that the wood you are buying in some way contributed to the death of great apes and other animals who became easy to find and kill because roads were built for other reasons. The Rescued Wood Bowl Company in Fort Collins, Colorado, is setting an example by using only rescued and recycled wood, wood that was on the way to a landfill.

There is also a move in the music world by the Gibson Guitar Company to use what is called "smartwood" to replace wood from dwindling rain forests. Who would have ever thought that playing a guitar may be directly related to killing animals, not to mention killing lovely trees and decimating fragile habitats?

Although many of us live far away from places where bushmeat is slaughtered and consumed, we can protest this illegal activity by being careful about what we buy and by expressing our outrage at this carnage. Each of us counts, and we all can make a difference if we act. Check out these websites:biosynergy.org/bushmeat/links.htm; members.aol.com/mjartisian/PrimatesAnimalsHope.html; send e-mail to: <bushmeat@biosynergy.org> to voice your opinions.

CHAPTER 12

ANIMALS, EYE SHADOW, AND FUR: USING ANIMALS FOR PRODUCT TESTING AND CLOTHING

In addition to being used for food, millions of animals, including dogs, cats, rats, mice, guinea pigs, and rabbits are used for testing non-essential cosmetic products such as deodorants, shampoos, soaps, and eye make-up (see Animalearn Fact Files, Product Testing). Neither the United States Food and Drug Administration nor the Consumer Product Safety Commission requires that animals be used to test the safety of products, but many companies still use animals for such purposes.

Lethal dose tests

Lethal dose tests try to measure the toxicity or potential harmfulness of products by using live animals. Animals receive a single dose of the substance to be tested either in their mouths, by stomach tubes, by inhaling a vapor powder or spray, by having it applied to the skin, or intravenously. The dose at which fifty percent of the animals die is called the Lethal Dose 50 (LD50) and that at which one hundred percent die is called the LD100. Many animals become sick and suffer greatly by experiencing convulsions, seizures, muscle cramps, abdominal pain, paralysis, and bleeding from the ears, eyes, nose, and rectum. If too many or too few animals die, the tests need to be repeated.

In addition to the inhumanity of these tests, the results are specific only to the conditions in which they were used, so they cannot be generalized from species to species or even between males and females of the same species. The LD50 is often used to estimate the safe dose of a given product for humans. For example, paraquat was introduced as an herbicide in 1960. Because the LD50 for rats was 120 milligrams per kilogram of body weight, it was thought that humans exposed to fewer milligrams of paraquat would be safe. However, in twelve years, more than four hundred humans died from exposure to this chemical, and it was estimated that the LD50 was much smaller, about four milligram per kilogram of body weight. Shockingly, over one hundred thousand people die annually from side effects of animal-tested drugs, and yet drugs and chemicals are still released out onto the market as "safe" because they have been tested by animals.

The Draize test

For a long time, rabbits have been used to test cosmetics. The late Henry Spira, founder of New York-based Animal Rights International, formed a Coalition to Abolish the Draize Test in 1979, and eventually achieved radical changes in product safety testing world-wide. (Spira's accomplishments in the field of animal protection are simply amazing and are well-documented in Peter Singer's book, *Ethics Into Action: Henry Spira and the Animal Rights Movement*.) The Draize test for eye irritation is named for a Federal Drug Administration scientist, John Draize, who standardized the scoring system of a pre-existing test for eye irritation in 1944. In the Draize test, a liquid or solid substance is placed in one eye of each rabbit in a group. The changes in the cornea, conjunctiva, and iris are then observed and scored. The rabbits' eyes are inspected at twenty-four, forty-eight, and seventy-two hours and at four and seven days. Both injury and potential for recovery are noted. The animals suffer immensely. Consumer protests against widespread use of the Draize test created the momentum which led to the development of non-animal alternatives to many types of animal testing. Henry Spira's campaign against the Draize test unleashed a growing movement against causing animals discomfort. By 1981, the cosmetics industry awarded one million dollars to Johns Hopkins School of Hygiene and Public Health to establish the Center for Alternatives to Animal Testing (CAAT). American Anti-Vivisection Society (AAVS) is involved in supporting the work of CAAT.

Testing environmental pollutants

Animals are also used in assessing the effects of environmental pollutants (toxicology research), poisons that cause cancer (carcinogens), birth defects, and numerous other diseases. Included in the group of toxic agents are trichloroethylene (TCE), used in spices, general anesthetics, and to decaffeinate coffee; lindane and dichloro-diphenyl-trichloro-ethane (DDT), two pesticides; and DES, a chemical that promotes growth in cattle and poultry. The negative influence of DDT on wildlife and the environment was the focus of Rachel Carson's famous and influential book *Silent Spring*, published in 1962. She noted how devastating the effects of DDT were and anguished over the loss of singing birds due to extensive poisoning; hence the title of her book.

The use of animals to study environmental poisons often produces highly questionable results. In many cases data from animals simply do not apply to humans in any direct way, and often the methods that are used are not adequate to draw meaningful conclusions about how pollutants affect humans. There are numerous technical and scientific problems centering on the use of animals to make predictions about human responses to drugs and environmental chemicals, many of which are related to the lack of progress science has made in helping humans deal with the negative effects of environmental poisons.

A good place to start reading in this area is Alix Fano's book, *Lethal Laws: Animal Testing, Human Health and Environmental Policy*. Fano reviews available data in the field of toxicology testing, shows why animal models rarely, if ever, work, and then suggests numerous non-animal alternatives that produce more reliable results. Some non-invasive, non-animal alternatives include the use of human cell cultures, human living tissues, computer-based models, and human volunteer studies. Some prestigious scientists agree with Fano and others who claim that the use of animals in toxicology testing is more the result of historical precedent than a careful assessment of the very limited utility of animal models. Fano quotes Philip Abelson, a famous American scientist and former editor of *Science*, a very prestigious scientific publication, as saying that, "The standard carcinogen tests that use rodents are an obsolescent relic of the ignorance of past decades." These tests do not work and should be stopped.

Some major criticisms of the use of animals to test the effects of potentially harmful chemicals on humans include the unusually large doses of poisons that are given to the animals and the fact that often the chemicals are administered in ways that do not mimic the human experience. In some studies, for example, hair dyes were *fed* to animals. Fano also points out that despite the fact that some animal tests show clear evidence of the negative effects of chemicals on the animals, the results have been ignored, and the chemical industry has actually grown rather than been slowed down.

Animals as clothing

How animals are used to manufacture clothes is also being scrutinized. Wild fur-bearing animals, over forty million individuals, are cruelly maimed and killed for profit (see Animalearn, Fact Files, Fur). Many are trapped using various types of contraptions that cause enormous psychological and physical suffering. These devices include leghold traps, wire snares that encircle an animal and pull tighter as the animal struggles, and conibears that grip the entire body and break the neck or back. Beavers are often trapped in water and drown after struggling for some time. Often companion dogs are accidentally trapped when traps are set to capture other species. There are no laws in the United States regarding how trapped animals can or cannot be killed.

Animals are also raised on farms (ranched-raised) only to be slaughtered for clothing. Recently it has been documented that dogs and cats have been used to make fur products. As are most animals raised for fur, these individuals typically are kept in deplorable conditions before being killed by being beaten, hanged, suffocated, or bled to death. In the United States, there are no federal laws prohibiting the import of dog or cat fur, or its use in clothing.

There also are no laws in the United States that regulate fur farms. Needless to say, farmed animals suffer from the same maladies as do captive animals in zoos. However, farmed animals are killed and are not even able to live out their lives in cages. The fur industry has a set of guidelines but its use is voluntary and there is no monitoring of fur farms. Animals such as mink are killed by neck-snapping ("popping"). They show great distress when removed from their cages to be killed - screeching, urinating, defecating - fighting for their lives. Gassing is also used, as are lethal injections, both of which cause pain and prolonged suffering before the animals are blessed with death. The carcasses of some farmed animals are even sold for dissection, so there is a connection between raising animals for fur and their use in education. Supporting dissection can also support the fur industry.

What can you do?

There are many things you can do about the plight of animals who are used for testing and clothing. Buy cruelty-free products that are not tested on animals. Have your parents buy stock in companies that practice cruelty-free testing of their products. Buy products that do not come from animals, rather than leather or fur. Boycott stores that sell fur. Do not dissect animals. Campaigns against fur-wearing have resulted in large declines in fur sales in the last decade, but many people still continue to wear fur and leather products. By going to the mass media and reaching millions of concerned consumers about the horrible treatment to which animals are subjected in the clothing industry, the lives of thousands of animals have been saved. Nonetheless, much more needs to be done.

Educate others. By setting an example yourself and by talking to others, you can truly help animals who cannot speak for themselves.

Other examples of animal models that do not work: Social deprivation and eating disorders

Clearly, many researchers believe little knowledge useful to humans has been compiled using animal models despite enormous investments of time, money, and animals' lives (see Animalearn Fact Files, Animal Experiments and also *Sacred Cows and Golden Geese*, by C. Ray and Jean Swingle Greek, a medical doctor and veterinarian). In the behavioral sciences, two examples of the inadequacy of animal models are the use of maternal and social deprivation to learn about human depression, and the use of animals to understand human eating disorders, including obesity, anorexia, and bulimia.

Following the work of Harry Harlow at the University of Wisconsin, socially deprived monkeys are commonly used to study psychological and physiological aspects of depression. Martin Stephens has published a valuable criticism of this work, as has the Animal Protection Institute (*Mainstream*, winter 1997, volume 28). Individuals typically are removed from their mothers and others soon after birth and raised alone, often in small, dark, and barren cages called "depression pits." In their impoverished prisons, isolated monkeys scream in despair, become self-destructive, and eventually withdraw from the world. The only social contacts with these unsocialized, frightened, and distraught monkeys occur when blood is drawn or other physiological measures are taken, or when they are introduced to other monkeys whom they avoid, or who maim or occasionally kill them.

Besides the fact that these types of studies are ethically repulsive, numerous flaws plague deprivation studies. Nonetheless, they are heavily funded by federal agencies as if both the lack of human clinical relevance and animals' lives do not matter. They are big business. But even researchers view human depression as a distinctly human condition. Simplistic animal models of human depression do not work for the diagnosis, treatment, or prevention of human depression. People who support other forms of animal use are offended by the horrors of deprivation research. Many believe it should be stopped immediately. No ends justify the means.

Kenneth Shapiro, in his book *Animal Models of Human Psychology: Critique of Science, Ethics and Policy,* has written extensively about the use of animal models in psychological research, specifically in eating disorders. He found that only thirty-seven percent of the clinicians who treat these conditions knew about research in which animals were starved, force-fed, or subjected to binge-purge cycles. Of those who did, eighty-seven percent said animal models were *not* used in treatment programs. The successful use of animal models for application in human clinical practice is extremely low. You probably would not go to the movies if you had the same slim chance of arriving successfully.

Convenience and tradition often drive animal use, but neither can adequately defend it, even in biomedical and toxicological research. Unfortunately, the use of animal models often creates false hopes for humans in need. It is estimated that only 1-3.5% of the decline in the rate of human mortality since 1900 has stemmed from animal research. The prestigious professional publication, the *New England Journal of Medicine*, called the war on cancer a qualified failure. And over one hundred thousand people die annually from side effects of animal-tested drugs.

Early animal models of polio also impeded progress on finding a vaccine. As pointed out by the Medical Research Modernization Committee, Dr. Simon Flexner's monkey model of polio misled other researchers concerning the mechanism of infection. He concluded that polio infected only the nervous systems of monkeys. However, research using human tissue culture showed that poliovirus could be cultivated on tissue that was not from the nervous system. Chimpanzees were used to study AIDS, but they do not contract AIDS. Countless humans die from some biomedical models because the diseases from which the animals suffered had to be artificially induced, and the course of the disease is different from naturally occurring conditions in humans.

Non-animal alternatives - including human studies that are more time-consuming, expensive (www.pcrm.org/issues/ Animal_ Experimentation_Issues/cost_analysis.html), risky, and more difficult to defend ethically than animal studies - need to be developed and used to learn about human behavioral and medical problems.

Not only will numerous humans benefit, but so will countless innocent animals.

Researcher effects in fieldwork:
We do make a difference

While there are many ethical concerns in studies of captive animals, there are also ethical issues associated with the study of wild animals. "Just being there" and walking around can have enormous impacts on the lives of wild animals. Nonetheless, field studies contribute information on the complexity and richness of animal lives that is very useful to those interested in animal well-being. Students of behavior often want to be able to identify individuals, assign gender and age, follow individuals as they move about, or record various physiological measurements including heart rate and body temperature. However, animals living under field conditions are generally more difficult to study than individuals living in more confined conditions, and, for instance, various methods (e.g. trapping, marking, fitting telemetric devices) are often used to make them more accessible. The use of leg-hold traps results in numerous serious injuries to animals, including swelling of limbs, lacerations, fractures, and amputation.

I have long been concerned with how various methods of study can influence the animals being studied (for example, their nesting and reproductive patterns, dominance relationships, mate choice, use of space, vulnerability to predators, feeding and care-giving behaviors). Models that are generated from these studies can be misleading because of human intrusions that appear to be neutral. We need to be sure that the behavior patterns documented are truly an indication of who the individual is in terms of such variables as age, gender, and social status. If the information used to make assessments of well-being are unreliable, then it follows that the conclusions that are reached and the animal models that are generated are also unreliable and can mislead current and future research programs. And, of course, our errors can have detrimental effects on the lives of the animals being studied. Our research ethic should require that we learn about the normal behavior and natural variation of various behavior patterns, so that we know just what we are doing to the animals we are studying.

Here are some representative studies to show how wide-spread researchers' influences are, and the diverse number of species that are affected by their influence.

Capturing and recapturing large grey mongooses influence their use of space, and even minor effects need to be considered in research projects involving such practices. It is important to compare their natural patterns of space use with their movements when they attempt to avoid traps or human observers. If, for example, cages are being designed to take into account animals' movement patterns, then data that are used to make decisions about designing enclosures need to be based on information that reliably indicates what the animals typically need in the wild.

Magpies, large and very intelligent birds who are not accustomed to human presence, spend so much time avoiding humans that this takes time away from essential activities such as feeding. If a researcher were interested in gathering data on feeding patterns by these birds, she would have to be sure that her presence did not change feeding patterns that are typical of the species.

Adélie penguins exposed both to aircraft and humans show profound changes in behavior, including deviation from a direct course back to a nest, and increased nest abandonment. Overall effects, due to exposure to aircraft that prevented foraging penguins from returning to their nests, included a decrease of fifteen percent in the number of birds in a colony and an eight percent increase in nest mortality rate. There were also large increases in penguins' heart rates. Here, models concerning reproductive success and parental investment would be misleading, once again because of the methods used.

Trumpeter swans do not show such adverse effects to aircraft. However, the noise and visible presence of stopped vehicles produces changes in incubation behavior by Trumpeter females that could result in decreased productivity due to increases in the mortality of eggs and hatchlings. Once again, data on the reproductive behavior of these birds would be misleading.

The foraging behavior of Little penguins (average mass of 1,100 grams) is influenced by their carrying a small device (about sixty grams) that measures the speed and depth of their dives. The small attachments result in decreased foraging efficiency. Changes in behavior such as these are called the "instrument effect." However, when female spotted hyenas wear radio collars weighing less that two percent of their body weight, there seems to be little effect on their behavior. Similar results have been found for small rodents, for whom small radio-collars do not increase the risk of predation by birds.

Placing a tag on the wing of ruddy ducks leads to decreased rates of courtship and more time sleeping and preening. In this case, data on activity rhythms and maintenance behaviors would have been misleading.

Mate choice in zebra finches is influenced by the color of the leg band used to mark individuals, and there may be other types of influence that have not been documented. Females with black rings and males with red rings have higher reproductive success than birds with other colors. Blue and green rings are especially unattractive on both females and males. Leg-ring color can also influence song tutor choice in birds such as zebra finches and mate-guarding in birds such as bluethroats.

The weight of radio-collars can influence dominance relationships in adult female meadow voles. When voles wear a collar greater than ten percent of their live body mass, there is a significant loss of dominance. Here, erroneous data concerning dominance relationships would be generated in the absence of this knowledge.

Methods of trapping can also lead to misleading results. Trapping methods can bias age ratios and sex ratios in some species of birds. Mist nets capture a higher proportion of juveniles, whereas traps capture more adults. Furthermore, dominant males tend to monopolize traps that are baited with food, leading to erroneous data on sex ratios. These are extremely important results because age and sex ratios are important data for many different researchers interested in behavior, behavioral ecology, and population biology.

Not only do research methods influence a wide variety of behavior patterns, but they can also influence susceptibility to infection. Ear-tagging white-footed mice leads to higher infestations by larval ticks because the tags impede grooming by these rodents. Thus, for researchers interested in grooming behavior, the presence of ear tags could influence results.

These examples suffice to show that animal models that do not take into account researcher effects can provide erroneous information about the behavior of numerous and diverse species. And if this misinformation is used to design future studies or captive housing or to evaluate well-being, then erroneous conclusions may be drawn.

Taking precautions

While we often cannot know about various aspects of the behavior of animals before we arrive in the field, our presence does seem to influence what animals do when we enter into their worlds. What appear to be relatively small changes at the individual level can have wide-ranging effects in both the short-and long-term. On-the-spot decisions often need to be made, and knowledge of what these changes will mean to the lives of the animals demands serious attention. A guiding principle should be that wild animals we are privileged to study should be respected, and when we are unsure about how our activities will influence their lives, we should err on the side of the animals and not engage in these practices until we know the consequences of our acts. This precautionary principle will serve us and the animals well.

The consideration of important ethical questions can be enriching in that we may come to consider new possibilities for how we interact with other animals, possibilities that enrich other animals as well as ourselves. There is a continuing need to develop and improve general guidelines for research on free-living and captive animals. Professional societies can play a significant role in the generation of guidelines, and many professional journals now require that contributors provide a statement saying that the research conducted was performed in agreement with accepted regulations. Guidelines should be progressive as well as regulatory. We should not be satisfied that things are better now than they were in the "bad old days," and we should work for a future in which even these enlightened times will be viewed as the bad old days for animals.

Much progress has already been made in the development of guidelines, and the challenge is to make them more binding, effective, and specific. I, along with Jane Goodall, am establishing a group called "Ethologists for the Ethical Treatment of Animals" to help ensure that the highest of ethical standards are enforced in behavioral research. If possible, we should also work for consistency among countries that share common attitudes towards animals; research in some countries (e.g. the United States) is less regulated than research in many other countries. Researchers who are exposed to the pertinent issues, and who think about them and engage in open and serious debate, can then carry these lessons into their research projects and import this knowledge to colleagues and students. Not knowing all of the subtleties of philosophical arguments, details over which even professional ethicists disagree, should not be a stumbling block nor an insurmountable barrier to learning.

CHAPTER 13

DISSECTION AND VIVISECTION: YOU DO NOT NEED TO CUT ANIMALS TO LEARN ABOUT LIFE

Alternatives to the use of animals:
The three R's - reduction, refinement, and replacement

Because of increasing pressure to reduce the use of animals in research and education, many people are interested in developing non-animal alternatives. The idea of the Three R's, *reduction, refinement, and replacement* of laboratory animal use, first appeared in a book published in 1959 titled *The Principles of Humane Experimental Technique,* written by two British scientists, William M. S. Russell and Rex Burch. This book was the first to present in a clear fashion how animals could be protected from human abuse. Many researchers use the Three R's today.

Reduction alternatives use fewer animals to obtain the same amount of data or allow more information to be obtained from a given number of animals. The goal of reduction alternatives is to decrease the total number of animals used.

Refinement alternatives lessen animal pain and distress. When developing refinement alternatives it is important to assess the level of pain an animal is experiencing. It is appropriate to assume that if a procedure is painful to humans, it will also be painful to animals. Refinement alternatives include the use of analgesics and/or anesthetics to alleviate any potential pain. Environmental enrichment, discussed above, is also an example of refinement.

Replacement alternatives are methods that do not use live animals, such as "in vitro" (meaning "in glass") systems. In vitro studies use living material or parts of living material cultured in petri dishes or in test tubes. "In vivo" studies are those carried out "in living animals." Mathematical and computer models can also be used to replace animals.

Are dissection and vivisection all they are cut out to be?

Opinions vary on whether it is essential to dissect dead animals or to vivisect - experiment on live animals - to learn about them. Members of about 170 species, including at least 10 million vertebrates are used annually for education in the United States. It has been estimated that about ninety percent of the animals used for dissection, including frogs, turtles, and fish, are wild caught. The philosopher Stephen Sapontzis has pointed out that killing and dissecting can teach bad attitudes towards animals and can lead students to think that animals are weak, that exploitation of the weak by the strong is permissible. Teachers and students can also become desensitized to the plight of other animals, and lose respect for the animals being used (see Animalearn, Fact Files, Animals in Education and Animal Experiments).

Many schools require students to cut apart or *dissect* dead animal specimens, or to do experiments on live animals, to *vivisect* them. Most students do not say anything to teachers about their objections to dissection or vivisection, and many students do not know nor are they told that there are many non-animal alternatives options that are readily available (for detailed information see the website of the American Anti-Vivisection Society, AAVS (www.aavs.org/). Ridicule, humiliation, lost time and perhaps feeling that they have to change career choices may make students decide to do something that they do not want to do. For example, Jonathan Balcombe, Associate Director for Education at the Humane Society of United States, noted that in 1995, all twenty-four county school systems in Maryland permitted students to use alternatives to dissection, but only one county had a written policy that required students and/or parents to be notified of this option.

Supporters of dissection frequently argue that "hands on" dissection or vivisection experience is essential to the student's education. Some biologists think that if someone does not want to cut animals up, he or she should not study biology. These scientists overlook the fact that there are many different types of biology that range from anatomical or physiological studies to watching animals behave.

There is also no evidence for the claim that "hands on" dissection or vivisection experience is essential to the student's education. Many teachers who use dissection and vivisection in their classrooms do so despite the fact that they do not know if exercises such as these actually work. Appeals to history, saying that this is the way we always did it, often provide weak reasons for continuing practices that either should never have been freely implemented in the first place, or for continuing practices that are simply outdated because of advances in other fields.

Along these lines, the Human Anatomy and Physiology Society (HAPS), in the absence of supporting data, states that "dissection and the manipulation of animal tissues and organs are essential elements in scientific investigation and introduce students to the excitement and challenge of future careers." While there is no doubt that dissection, vivisection, and experimentation historically have played major roles in many courses, it is not at all clear that these activities "are essential" in the sense that science could not be carried on without them, particularly today with all of the technological advances. Furthermore, HAPS's view is not necessarily consistent with the "nature of scientific inquiry" as practiced today. There are many different types of inquiry that can be labeled "scientific," and there seems to be more open-mindedness even among practicing scientists. Certainly, it would undermine educating future scientists if they were taught that there was only one way to teach science, including anatomy and physiology. Furthermore, defining "science" as dissection or vivisection is misleading, for most scientific endeavors, broadly defined, do not involve these practices.

A recent survey showed that more than half of American medical schools do not currently use live animals in their regular medical curricula, and of those that do, 125 out of 126 offer alternative exercises for students who do not want to participate directly in procedures that use live animals. Thus, almost all medical schools allow students to graduate if they have done no surgery or other laboratory exercises using live animals. There is increasing discussion of these issues in veterinary schools because more and more students are becoming interested in animal rights and animal use in veterinary education, and some schools already offer non-animal alternatives to dissection and vivisection. This information is regularly updated on the website for the Association of Veterinarians for Animal Rights, AVAR (www.avar.org/).

The educational effectiveness of non-animal alternatives

There are many studies comparing the educational effectiveness of alternatives, such as computer software and models, and they show that alternatives often are at least as good if not better for achieving intended educational goals. Jonathan Balcombe summarized some of these and found that for undergraduates, veterinary students, and medical students, equal knowledge or equivalent surgical skills were acquired using alternatives. The educational effectiveness of non-animal models was as good if not better than animal models (www.hsus.org/). For example, in a study of 2,913 first-year biology undergraduates, the examination results of 308 students who studied model rats were the same as those of 2,605 students who dissected rats. When the surgical skills of thirty-six third-year veterinary students who trained on soft-tissue organ models were compared to the surgical skills of students who trained on dogs and cats, the performance of each group was the same. Virtual surgery has been shown to be an effective alternative. In a study of 110 medical students, students rated computer demonstrations higher for learning about cardiovascular physiology than demonstrations using dogs. Much information about non-animal alternatives can be found at these websites: www.mindlab.msu.edu; www.avar.org/; www.pcrm.org/

What can you do?

You *can* request non-animal alternatives if you do not want to dissect a live earthworm, pith a frog, or work on already prepared specimens like cats or fetal pigs. A valid alternative is one that harms no animals. Watching others dissect is not necessarily acceptable; the alternative is one that involves no contact, either direct or indirect, with animals. Studies show that using alternatives educates as well or better than working on animals, and additional alternatives to animal use are continually being developed.

There also are some guidelines that might be useful for you or others who are helping you along. Gary Francione and Anna Charlton, in their book, *Vivisection and Dissection in the Classroom: A Guide to Conscientious Objection,* list eight steps that are useful to follow before taking legal action. These include: know how far you are willing to go to assert your right not to engage in vivisection or dissection; raise your objection at the earliest time; approach your teacher alone or with like-minded classmates and be prepared to discuss why you object to vivisection or dissection; assess the situation carefully and intelligently; be prepared to present one or more alternatives; document everything; and seek legal help early and organize your network of support.

Finding non-animal alternatives

Many students seek out non-animal alternatives.
(see www.hsus.org/programs/research/annotate.html).
There are various ways to find out about alternatives (see
www.aavs.org/animalearn and the Animals in Education fact sheet
published by AAVS). A useful source of alternatives can be obtained
from the database known as *NORINA* (A Norwegian Inventory
of Audiovisuals; oslovet.veths.no/NORINA). Since 1991,
information on over 3,500 audiovisual aids (and their suppliers)
has been collected that may be used as animal alternatives or
supplements in the biomedical sciences at all levels from primary
schools to University. Other sources for finding out about non-
animal alternatives include my *Encyclopedia of Animal Rights and
Animal Welfare,* Amy Blount Achor's *Animal Rights: A Beginner's
Guide,* and Ursula Zinko, Nick Jukes, and Corina Gericke's *From
Guinea Pig to Computer Mouse: Alternative Methods for a Humane
Education.* There are centers for the development of alternatives to
animals at Johns Hopkins University and the University of California
at Davis, and they publish newsletters and other material concerning
alternatives. You can also call Animalearn at 1-800-SAY-AAVS for
more information on any issue concerning dissection or vivisection
in the classroom.

It is all right to question how science is taught; In fact, it is important to do so!

It is important for you to recognize that questioning how science
is taught is not to be against science, anti-intellectual, or "radical."
Indeed, one could argue that HAPS's assertion that "dissection and the
manipulation of animal tissues and organs are essential elements
in scientific investigation and introduce students to the excitement and
challenge of future careers" in the absence of supporting data is
anti-scientific. Questioning science will make for better, more
responsible science and education.

Decisions against using animals do not compromise sound
education. Resistance against the use of alternatives because they do
not work is more dogma than fact. Dissection and vivisection are not
all they are cut out to be. It is not essential to cut or kill animals to
learn about life. There are always alternatives to cruelty.

CHAPTER 14

WHERE TO GO FROM HERE?
WE ARE THE KEY TO THE FUTURE

Care, share, show respect, be compassionate, be tolerant, talk to other animals and listen to them, and tread lightly

It is all too easy to talk about caring about other animals and the environment and then to walk away from the problems at hand because there will always be others who are not as busy and who will be concerned and will act. It is too easy to pass accountability *and* responsibility on to others.

So, what do we need to do? It is simple - we need to ask why we continue to do the horrible things that we do to animals, and why we ruin the environment. It is very clear that what seems to have worked in the past really has not worked at all. Some very big changes have to occur in the very near future, not when it is convenient to make them. *Time is not on our side, there is a serious sense of urgency.*

Roger Fouts, a psychologist who has studied chimpanzee-human communication for a long time with individual chimpanzees, including Washoe, recently wrote the wonderful book *Next of Kin*. In this book, Fouts concluded that by being concerned about animals and acting on these concerns, you are more of a healer than an activist.

Although not all humans feel that animals or the environment are here for us to use and abuse selfishly, human arrogance prevails in many circles, and animals suffer great losses and immeasurable harm because of human-centered attitudes and domination of living (animate) and non-living (inanimate) environments. To be sure, in order to save the precious and fragile resources on this planet and in the universe at large, humans will have to place their own anthropocentric, selfish interests aside and work with, and not against, one another. Humans will also have to work in harmony with, and appreciate and respect the value of other living and non-living cohabitants of this planet and of the universe. They will have to make serious attempts to imagine what it is like to be the animals whose lives they are compromising. *And their feelings will have to be incorporated into action.*

Getting out and doing something: Each of us counts, each of us makes a difference

There is so much you can do to help other animals and the world as a whole. This is a wonderful world, and nature is so very generous with her gifts. As I said before, there are *always* alternatives to cruelty, and it is important to remember this.

Maintaining close contact with pro-animal organizations such as AAVS (www.aavs.org/) is a good place to start. Jane Goodall developed the Roots & Shoots program that requires groups of young people to participate in projects that benefit the environment, animals and the human community in their areas. The program has two major messages: (a) that every individual matters and can make a difference, every day; and (b) "Only when we understand can we care; Only when we care shall we help; Only if we help will all be saved" (see www.janegoodall.org/; www.janegoodall.org/rs/rs_history.html). You can also contact the head of pain and distress campaign at the Humane Society of the United States (www.hsus.org).

We need to talk among ourselves and connect with other animals. We must include animals in our discussions about their lives. Forming bonds with other animals helps in forming bonds with other humans. Contact of young humans with companion animals is associated with children's social development, including their feelings for other children. Green Chimneys, an organization dedicated to providing care and concern for all living things, also exposes people to the benefits of developing close ties with animals (www.pcnet.com/ ~gchimney/index.html). We all hold the key to the future. We are the future. Listen to your heart when it tells you that things are going wrong.

In addition to asking questions such as the ones I have presented, there are other things that you can do. Here are some ideas. I am sure that you can add many more to this list.

How animals are represented - it makes a difference

How animal images and live animals are represented in advertisements, on television, in movies, and in cartoons also influences what people come to believe about them. Often they are given human characteristics, especially the ability to talk. They are dressed in clothes and adorned with wigs and make-up. The Chimpanzee Channel misleadingly portrayed individuals in this manner, and many people are concerned about how chimpanzees and orangutans are misrepresented as humans. They are not humans, of course, and should not be presented in this way.

The image of dolphins has also suffered at the hands of humans. Consider the case of Flipper, the famous dolphin (actually there were five Flippers) whose image was created to serve human ends. Richard O'Barry, the first trainer of Flipper, even believed that Flipper was an illusion, a fabrication of hundreds of people who created his legend. O'Barry's story, reported in the book *Behind the Dolphin Smile*, is an excellent example of someone who came to realize that keeping dolphins in tanks for commercial exploitation was wrong. The mentality that we can do whatever we want with such amazing animals as dolphins is what led to the exploitation of the image of Flipper and the development of large and very profitable businesses - "dolphinaria" - each with its obligatory Flipper. In the United Kingdom alone, it is estimated that about three hundred bottle-nose dolphins were imported, with more than one hundred deaths in captivity.

Recently, Elizabeth Paul, at the University of Edinburgh, showed that television programs in Great Britain supported the notion of a hierarchy of "higher" and "lower" animals, with lower animals perceived as not suffering as much as higher animals. Cruelty to mammals was not tolerated, but cruelty to fish and invertebrates was. Dr. Paul also found that mammals tended not to be shown as meat for meals, and she believes that this is related to adults being uncomfortable with advocating kindness to animals and then allowing them to be killed for food. They did not want their children to know that a hamburger is a cow on a bun. Nonetheless, it is important to spread this information widely.

When an animal is shown in a setting that is unrelated to its natural environment, a false message is presented. This prevents communicating an accurate understanding of the animal's nature. TV commercials often include mountain lions or other animals to show that a sofa is soft and comfortable. Because the lion is shown out of context as a soft and cuddly creature, it does nothing to promote an understanding or appreciation of the true characteristics of lions - who they are and how they live. These distortions also convey a false picture of human's place in nature and can have harmful effects in the future.

Try to take the animal's point of view

Portraying animals as objects promotes the view of animals as commodities. When something goes wrong, animals are too easily discarded or replaced. Try to take the animals' points of view. Imagine what their worlds are like to them. What it is like to be a bat, flying around, resting upside down, and having very sensitive hearing. Or what it is like to be a dog with a very sensitive nose and ears. Imagine what it is like to a free-running gazelle, or a wolf, coyote, or deer out in nature. Take advantage of what animals selflessly and generously offer to us. Their worlds are truly awe-inspiring. It is essential not only to *look* at animals but to *see* them, to *see* them as they actually are, not as we want them to be.

The field biologist, David Macdonald, at Oxford University in England, is proud to exclaim: "I study foxes because I am still awed by their extraordinary beauty, because they outwit me, because they keep the wind and the rain on my face . . . because it is fun." It should be fun to make animals' lives better! Now, imagine what it would be like to be caged, trapped, tagged, handled, confined, restrained, isolated, mutilated, shocked, starved, deprived, dropped in hot water, or unable to escape from shock in an experiment. It is not a pretty picture.

Is it truly believable that humans are the only species that can think, feel pain, experience anxiety, and suffer? Even if we are very different from dogs or cats, there is no reason to think that dogs, cats, and many other animals do not think and feel pain and suffer in their own ways. Many people who think that *only* humans are conscious or sentient are really trying to protect their own high position on scales of nature in which humans are placed above and apart from other animals.

As Richard Ryder, the British psychologist turned animal

advocate, has noted in his book *The Political Animal: The Conquest of Speciesism*: "The simple truth is that we exploit the other animals and cause them suffering because we are more powerful than they are. Does that mean that if aliens land on Earth and turn out to be far more powerful than us we would let them, without argument, chase and kill us for sport, experiment on us or breed us in factory farms and turn us into tasty humanburgers? Would we accept their explanation that it is perfectly moral for them to do all these things because we are not members of their species?"

Humans are a part of nature, not apart from nature: Bringing animals into our hearts and loving nature

Thomas Berry stresses that we should strive to have a benign presence in nature. I agree. *Humans are a part of nature, not apart from nature*. Humans cannot continue to be at war with the rest of the world. The fragility of the natural order - the delicate balance of life - requires that we all work harmoniously so as not to destroy nature's wholeness, goodness, and generosity.

A compassionate caring and sharing ethic is needed now so that the interconnectivity and spirit of the world will not be lost. In the absence of animals, we would live in a severely impoverished, unstimulating universe. How sad this would be.

Expanding our circle of respect and understanding can help bring us all together. The community "out there" needs to become the community "in here" - in our hearts. Feelings need to be incorporated into action. *We need to love nature.*

Ethical enrichment

It is in the best traditions of science to ask questions about ethics. Ethics can enrich our views of other animals in their own worlds and in our different worlds, and help us to see that variations among animals are worthy of respect, admiration, and appreciation. The study of ethics can also broaden the range of possible ways in which we interact with other animals without ruining their lives. Ethical discussion can help us to see alternatives to past actions that have disrespected other animals and, in the end, have not served us or other animals well. In this way, the study of ethics is enriching to other animals and to ourselves. The consideration of important ethical questions can be enriching rather than stifling in that we may come to consider new possibilities for how we interact with other animals, possibilities that enrich other animals as well as ourselves. If we think that ethical considerations are stifling and create unnecessary hurdles over which we must jump in order to get done what we want to get done, then we will lose rich opportunities to learn more about other animals and also ourselves. *Our greatest discoveries come when our ethical relationships with other animals are respectful and not exploitative.*

The separation of "us" from "them" presents a false dichotomy, the result of which is a distancing that erodes rather than enriches the possible numerous and intimate relationships that can develop among all animal life. As the primatologist, Barbara Smuts at the University of Michigan, wrote: "My own life has convinced me that the limitations most of us encounter in our relations with other animals reflect not their shortcomings, as we so often assume, but our own narrow views about who they are and the kinds of relationships we can have with them."

Doing science and respecting animals

Studying non-human animals is a privilege that must not be abused. We must take this privilege seriously. Although the issues are very difficult and challenging, this does not mean it is impossible to deal with them. *Certainly we cannot let the animals suffer because of our inability to come to terms with difficult issues.* Questioning the ways in which humans use animals will make for more informed decisions about animal use. By making such decisions in a responsible way, we can help to insure that in the future we will not repeat the mistakes of the past, and that we will move towards a world in which humans and other animals may be able to share peaceably the resources of a finite planet.

Thomas Dunlap, in his book *Saving America's Wildlife,* notes that the role of science in the development and changing of ideas is highly questionable. While many people respect scientists and bestow on them special abilities to fix things when they break, scientists and science alone will not be able to deal effectively with the many difficult and puzzling problems that arise in discussions of the nature of animal-human interactions. Personal and cultural values influence the choices we all make, and common sense also plays a large role in our decision making. In the future, "science" will have to incorporate values and facts, and scientists will have to figure out how they are to be factored into the choices we make.

Deep ethology: Respecting nature's ambassadors

I use the term "deep ethology" to stress that people recognize that they are not only an integral part of nature, but also that they have unique responsibilities to nature. "Deep ethology" also means respecting, appreciating, and showing compassion for all animals, feeling for all animals from one's heart. "Deep ethology" means resisting speciesism. A deep respect for animals does not mean that just because animals are respected we can then do whatever we want to them.

Our starting point should be that we will not intrude on other animals' lives unless we can argue that we have a right to override this maxim, that our actions are in the best interests of the animals, irrespective of our desires. When unsure about how we influence the lives of other animals, we should err on the side of the animals.

As I mentioned before, some guiding principles include: (1) putting respect, compassion, and admiration for other animals first and foremost; (2) taking seriously the animals' points of view; (3) erring on the animals' side when uncertain about their feeling pain or suffering; (4) recognizing that almost all of the methods that are used to study animals, even in the field, are intrusions on their lives (much research is exploitive); (5) recognizing how misguided are speciesistic views concerning vague notions such as intelligence and cognitive or mental complexity for informing assessments of well-being; (6) focusing on the importance of individuals; (7) appreciating individual variation and the diversity of the lives of different individuals in the worlds within which they live; (8) appealing to what some call questionable practices that have no place in science, such as the use of common sense and empathy; and (9) using broadly based rules of fidelity and non-intervention as guiding principles.

Forming friendships and partnerships with other animals: Loving the earth

If we want children to flourish, to become truly empowered, then let us allow them to love the earth before we ask them to save it. Perhaps this is what Thoreau had in mind when he said 'the more slowly trees grow at first, the sounder they are at the core, and I think the same is true of human beings.' (David Sobel. *Beyond ecophobia: Reclaiming the heart in nature education*)

We and all animals with whom we share our time and space

should be viewed as friends and partners in a joint venture. It is important to remember that there can be close connections among seemingly unrelated activities. Remember how logging and killing animals for bushmeat were related because hunters were able to use the roads that logging companies built to gain access to the animals they wanted to kill. Remember also how playing a guitar could be associated with killing animals and decimating fragile forests.

In our efforts to learn more about the worlds of other animals, we need to study many different species. We must not be afraid of what learning about other species may bring in terms of knowledge about animal consciousness, intelligence, and their ability to feel pain and to suffer. We cannot continue to view animal suffering from afar, nor should we blind ourselves to the many ways in which we cause harm to the world around us. We have become so accustomed to the way we do things that somehow we think that we are doing just fine.

Sue Savage-Rumbaugh, who is well-known for her studies of ape language with Kanzi, a bonobo, has stressed that it is time to change course, that we need to open our eyes, our ears, our minds, our hearts. We need to *look* with a new and deeper vision, to *listen* with new and more sensitive ears. It is time to *learn* what animals are really saying to us and to each other. These three L's should be used to get us to act on behalf of all animals. No one can be an island in this intimately connected universe.

What I fear most is that if we stall in our efforts to take animal use and abuse more seriously and fail to adopt extremely restrictive guidelines and laws, even more and irreversible damage will result. Our collective regrets about what we failed to do for protecting animals' rights in the past will be moot. We need to enter into close and reciprocal relationships with all beings in this more-than-human world. *We need to respect and love them.*

I hope that all animals on this planet benefit from open discussions about animal-human relationships. *When animals lose, we all lose. Every single loss diminishes us and the planet.* So when you are not sure about what is happening, feel free to talk to practicing scientists and those who use animals in other ways. Also, talk to your friends and relatives and tell them what you are feeling - it is never too late to make a change. Who knows, maybe you will all learn something that you can share with others.

While going to zoos or going hunting and fishing may be social events, and often family-oriented activities in which parents bond with their children, are there not better ways to spend family time than watching animals in cages or scaring, harassing, maiming, or killing them? Think about other activities in which you and your family and friends can partake, activities that are friendly to, and truly help, animals and the planet.

We are the voices for voiceless animals: Back to the ABC's

Remember, we are the voices for voiceless animals. Let all animals know us by our actions to take their lives seriously. Nature is generous, so let us return the favor. As in most aspects of life, you will receive what you give. If you give love and respect, you will receive love and respect.

So, I end as I started. **Always Be Caring and Sharing** - that is what the **ABC'S** of animal protection and compassion are all about. Making decisions about who lives and who suffers or dies is serious business. Speak out against animal abuse in zoos, circuses, rodeos, sport-hunting, factory farming, various types of research, dissection and vivisection, on television and in movies, and advertising. Boycott events in which animals are abused.

Be proactive. Prevent animal abuse before it starts. Henry Spira was able to make great strides towards abolishing the Draize test, and public pressure greatly influenced the use of veal. Sears, Roebuck, and Company recently stopped sponsoring the Ringling Brothers' circus. Two people, Helen Steel and Dave Morris, were able to take on McDonald's in the longest trial in British history - called the McLibel case - and show that McDonald's exploits children with their advertising, falsely advertises their food as nutritious, risks the health of their long-term regular customers, and is "culpably responsible" for cruelty to animals reared for their products (see: www.mcspotlight. org/case/trial/verdict/index. html and www.mcspotlight.org).

Often, the greater our ignorance about something, the greater our resistance to change. Sometimes we are afraid of the unknown. I hope that you now know more than you did before reading this book about how to protect and speak for other animals, and that you are able to make more responsible decisions after discussing the issues with family, friends, and teachers.

Remember, you hold the key to the future. You are the future. With knowledge comes growth. Each and every one of us makes a difference. We must stroll with our kin, not walk away from them.

TWELVE MILLENNIAL MANTRAS
BY MARC BEKOFF AND JANE GOODALL

The millennium is here. Let us take stock of who we are and where we are going. Is it acceptable to weep not only for human suffering but also for the rampant misery of other animals with whom we share the planet? Can we shed tears for Sissy, the severely beaten elephant at the El Paso Zoo, the kicked and abused elephants and chimpanzee, Trudy, at the Chippenfield Circus in England? Can we also weep for the millions of animals in laboratory prisons, the billions of animals tortured and slaughtered for food and clothing? Can we sincerely mourn the destruction of the natural world, the vanishing forests, wetlands, savannas and bodies of water? We hope these twelve mantras will make a difference for future generations:

One: Compassion and empathy for animals beget compassion and empathy for humans. Cruelty towards animals begets cruelty toward humans.

Two: All life has value and should be respected. Every animal owns her or his own life spark. Animals are not owned as property. All living creatures deserve these basic rights: the right to life, freedom from torture, and liberty to express their individual natures. Many law schools offer courses in animal law. If we agree, we would interact with animals in rather different ways. We shall need compelling reasons for denying these rights and ask forgiveness for any animal we harm.

Three: Do unto others, as you would have them do unto you. Imagine what it would be like to be caged, trapped, restrained, isolated, mutilated, shocked, starved, socially deprived, hung upside down awaiting death or watching others slaughtered. Biological data clearly show that many animals suffer physically and psychologically and feel pain.

Four: Dominion does not mean domination. We hold dominion over animals only because of our powerful and ubiquitous intellect. Not because we are morally superior. Not because we have a "right" to exploit those who cannot defend themselves. Let us use our brain to move towards compassion and away from cruelty, to feel empathy rather than cold indifference, to feel animals' pain in our hearts.

Five: Human beings are a part of the animal kingdom not apart from it. The separation of "us" from "them" creates a false picture and is responsible for much suffering. It is part of the in-group/out-group mentality that leads to human oppression of the weak by the strong as in ethnic, religious, political and social conflicts. Let us open our hearts to two-way relationships with other animals, each giving and receiving. This brings pure and uncomplicated joy.

Six: Imagine a world without animals. No birdsong, no droning of nectar searching bees, no coyotes howling, no thundering of hooves on the plains. Rachel Carson chilled our hearts with thoughts of the silent spring. Now we face the prospect of silent summers, falls and winters.

Seven: Tread lightly. Only interfere when it will be in the best interests of the animals. Imagine a world where we truly respect and admire animals, feel heart-felt empathy, compassion and understanding. Imagine how we should be freed of guilt, conscious or unconscious.

Eight: Make ethical choices in what we buy, do and watch. In a consumer-driven society our individual choices, used collectively for the good of animals and nature, can change the world faster than laws.

Nine: Have the courage of conviction. Never say never. Act now. Be proactive; prevent animal abuse before it starts. Dare to speak out to save the world's precious and fragile resources. Live as much as possible in harmony with nature, respecting the intrinsic value of all life and the wondrous composition of earth, water and air.

Ten: Every individual matters and has a role to play. Our actions make a difference. Public pressure has been responsible for much social change, including more humane treatment of animals. "Whistle blowers" have courageously revealed intolerable conditions in laboratories, circuses, slaughterhouses and so on, often at the expense of their jobs: Henry Spira organized peaceful demonstrations that led to questioning the Draize test in which rabbits are harmed to learn about the effects of eyeshadow.
His efforts also led to the formation of centers devoted to the development of non-animal alternatives, sponsored by the cosmetic companies themselves. Public pressure greatly reduced veal consumption and led to Sears, Roebuck, and Company ending their sponsorship of Ringling Brothers Barnum & Bailey Circus. Helen Steel and Dave Morris took on McDonald's in the longest trial in British history (the McLibel case) and showed that they exploit children with their advertising and are "culpably responsible" for cruelty to animals.

Eleven: Be a passionate visionary, a courageous crusader. Combat cruelty and catalyze compassion. Do not fear to express love. Do not fear to be too generous or too kind. Above all, understand that there are many reasons to remain optimistic even when things seem grim. Let us harness the indomitable human spirit. Together we can make this a better world for all living organisms. We must, for our children, and theirs. We must stroll with our kin, not walk away from them.

A millennial mantra: When animals lose, we all lose. Every single loss diminishes us as well as the magnificent world in which we live together.

Acknowledgements

I thank Katherine Lewis, Dodi Boone, and Stephanie Shain at AAVS for their detailed and very helpful comments on this manuscript. Sara Goering, Gary Francione, Margaret Wallace, Carron Meaney, Marjorie Bekoff, Barbara Fiedler, Bruce Gottlieb, Marta Turnbull, Bernie Rollin, Dale Jamieson, Colin Allen, Susan Townsend, and Jeff Masson also read an early draft of this book and/or discussed with me many of the issues considered herein

Resources

Here is a list of general resource material covering many of the issues discussed. This list is not meant to be complete but rather includes books and articles that are frequently cited, and journals that publish essays that are concerned with animal rights, environmental ethics, and animal behavior and cognition. Numerous references and organizations that deal with animal protection can be found in my *Encyclopedia of Animal Rights and Animal Welfare,* in Amy Blount Achor's *Animal Rights: A Beginners Guide,* at the website for *Animals' Agenda* (www.animalsagenda.org/DirOrgLinks.asp?menu=DirOrgLinks), and in the American Anti-Vivisection Society's (www.aavs.org/) Animalearn Fact Files.

Books, reports, and pamphlets

Abram, D. 1996. *The spell of the sensuous: Perception and language in a more-than-human world.* Pantheon Books, New York.

Achor, A. B. 1996. *Animal rights: A beginner's guide.* WriteWare, Yellow Springs, Ohio.

Adams, C. J. 1994. *Neither man nor beast: Feminism and the defense of animals.* Continuum, New York.

Adams, C. J. 2000. *The inner art of vegetarianism.* Lantern Books, New York, New York.

Allen, C. 1998. Assessing animal cognition: Ethological and philosophical perspectives. *Journal of Animal Science* 76, 42-47.

Allen, C. and Bekoff, M. 1997. *Species of mind: The philosophy and biology of cognitive ethology.* MIT Press, Cambridge, Massachusetts.

Animal Welfare Information Center Newsletter. 1995. Directory of resources on alternatives and animal use in the life sciences. United States Department of Agriculture, Beltsville, Maryland.

Animal Welfare Information Center Newsletter. 1996. A review of the animal welfare enforcement report data 1973-1995. United States Department of Agriculture, Beltsville, Maryland.

Animal Welfare Institute. 1990. *Animals and their legal rights.* Animal Welfare Institute, Washington, D. C.

Arluke, A. 1993. Trapped in a guilt cage. *Animal Welfare Information Center Newsletter* 4: 1-8.

Baird, R. M. and Rosenbaum, S. E. (editors) 1991. *Animal experimentation: The moral issues.*

Prometheus Books, Buffalo, New York.

Balcombe, J. 1997. Student/teacher conflict regarding animal dissection. *The American Biology Teacher* 59, 22-25.

Balcombe, J. 2000. *The use of animals in higher education: Problems, alternatives, and recommendations.* HSUS, Washington, DC.

Balls, M. et al. 1995. *The Three Rs: The way forward: The Report and Recommendations of ECVAM Workshop (European Centre for the Validation of Alternative Methods).* ATLA 23, 838-866.

Bateson, P. P. G. 1997. The behavioural and physiological effects of culling red deer. Report to the Council of the National Trust, United Kingdom.

Bayne, K. A. L. and Kreger, M. D. (editors) 1995. *Wildlife mammals as research models: in the laboratory and field.* Scientists Center for Animal Welfare, Greenbelt, Maryland.

Bekoff, M. 1994. Cognitive ethology and the treatment of non-human animals: How matters of mind inform matters of welfare. *Animal Welfare* 3, 75-96.

Bekoff, M. 1995. Marking, trapping, and manipulating animals: Some methodological and ethical considerations. In K. A. L. Bayne and M. D. Kreger (Eds.), *Wildlife mammals as research models: In the laboratory and field.* Scientists Center for Animal Welfare, Greenbelt, Maryland. pp. 31-47.

Bekoff, M. 1996. Are dissection and tissue and organ manipulation really essential? *Strategies* Spring 1996, 4-5.

Bekoff, M. 1998. Deep ethology. *AV Magazine* (A publication of the American Anti-Vivisection Society). 106 (1): pp. 10-18 (Winter 1998)

Bekoff, M. (editor) 1998. *Encyclopedia of animal rights and animal welfare.* (Foreword by Jane Goodall), Greenwood Publishing Group, Westport, Connecticut.

Bekoff, M. 1998. Resisting speciesism and expanding the community of equals. *BioScience,* 48, 638-641.

Bekoff, M. 1998. Deep ethology, animal rights, and the Great Ape/Animal Project: Resisting speciesism and expanding the community of equals. *Journal of Agricultural and Environmental Ethics,* 10, 269-296.

Bekoff, M. (editor) 2000. *The smile of a dolphin: Remarkable accounts of animal emotions.* Random House/Discovery Books, New York.

Bekoff, M. 2000. Human-carnivore interactions: Adopting proactive strategies for complex problems. In Gittleman, J. L., Funk, S. M., Macdonald, D. W., Wayne, R. K. (editors) *Carnivore conservation.* Cambridge University Press, London.

Bekoff, M. and Jamieson, D. 1996. Ethics and the study of carnivores. In Gittleman, J. L. (editor) *Carnivore Behavior, Ecology, and Evolution.* Cornell University Press, Ithaca, New York. pp. 16-45.

Berry, T. 1999. *The great work: Our way into the future.* Bell Tower, New York.

Bostock, S. St.C. 1993. *Zoos and animal rights.* Routledge, London.

Bowen-Jones, E. and Pendry, S. 1999. The threat to primates and other mammals from the bushmeat trade in Africa, and how this threat could be diminished. *Oryx* 33, 233-246.

Callicott, J. B. 1980/1989 Animal liberation: A triangular affair. In Callicott J. B. *In defense of the land ethic: essays in environmental philosophy.* State University of New York Press, Albany. pp. 15-38.

Campbell, T. Colin and C. J. Chen. 1994. Diet and chronic degenerative diseases: Perspectives from China. *American Journal of Clinical Nutrition* 59, 1153-61.

Carson, R. 1962. *Silent spring.* Houghton-Mifflin, Boston.

Cavalieri, P. and Singer, P. (editors) 1993. *The Great Ape Project: Equality beyond humanity.* Fourth Estate, London.

Collard, S.B. III. 2000. *Acting for nature: What young people around the world have done to protect the environment.* Heyday Books, Berkeley, California.

Crick, F. 1994. *The astonishing hypothesis: The scientific search for soul.* Scribners, New York.

Croke, V. 1997. *The modern ark: The story of zoos: Past, present, and future.* Scribners, New York.

Davis, H. and Balfour, D. (editors) 1992. *The inevitable bond: Examining scientist-animal interactions.* Cambridge University Press, New York.

Davis, K. 1996. *Poisoned chickens, poisoned eggs: An inside look at the modern poultry industry.* Book Publishing Company, Summertown, Tennessee.

Davis, S. G. 1997. *Spectacular nature: Corporate culture and the sea world experience.* University of California Press, Berkeley.

Dawkins, M. S. 1980. *Animal suffering: The science of animal welfare.* Routledge, Chapman and Hall, New York.

Dawkins, M. S. 1993. *Through our eyes only?* W H Freeman, San Francisco.

DeGrazia, D. 1996. *Taking animals seriously: Mental life and moral status.* Cambridge University Press, New York.

DeGrazia, D. 1999. Animal ethics around the turn of the twenty-first century. *Journal of Agricultural and Environmental Ethics* 11, 111-129.

Dickinson, P. 1988. *Eva.* Dell, New York.

Dol, M., Kasamoentalib, S., Lijmbach, S., Rivas, E. & van den Bos, R. (editors), *Animal consciousness and animal ethics.* Van Gorcum, Assen (The Netherlands)

Drayer, M. E. (editor) 1997. *The animal dealers: Evidence of abuse of animals in the commercial trade 1952-1997.* Animal Welfare Institute, Washington, D. C.

Duda, M. D., Bissell, S. J., and Young, K. C. 1996. Factors related to hunting and fishing participation in the United States. *Transactions of the 61st American Wildlife and Natural Resources Conference,* pages 324-337.

Dunlap, T. R. 1988. *Saving America's wildlife: Ecology and the American mind.* Princeton University Press, Princeton, New Jersey.

Eisnitz, G. A.1997. *Slaughterhouse: The shocking story of greed, neglect, and inhumane treatment inside the U. S. meat industry.* Prometheus Books, Buffalo, New York.

Fano, A. 1997. *Lethal laws: Animal testing, human health and environmental policy.* Zed Books, London.

Finsen, L. and Finsen, A. 1994. *The animal rights movement in American: From compassion to respect.* Twayne Publishers, New York.

Fouts, R. with S. Mills. 1997. *Next of kin: What chimpanzees have taught me about who we are.* William Morrow and Company, New York.

Fox, M. A. 1998. *Vegetarianism.* In Bekoff 1998.

Fox, M. A. 1999. *Deep vegetarianism.* Temple University Press, Philadelphia.

Fox, M. W. 1980. *Returning to Eden: Animal rights and human responsibility.* The Viking, Press, New York.

Fox, M. W. 1990. *Inhumane society: The American way of exploiting animals.* St. Martin's Press, New York.

Fox, M. W. 1992. *Superpigs and wondercorn.* Lyons and Burford, New York.

Fox, M. W. 1997. *Eating with conscience: The bioethics of food*. NewSage Press, Troutdale, Oregon.

Fox, M. W. 1999. *Beyond evolution: The genetically altered future of plants, animals, the earth . . . and humans*. The Lyons Press, New York.

Francione, G. L. 1995. *Animals, property, and the law*. Temple University Press, Philadelphia.

Francione, G. L. 1996. *Rain without thunder: The ideology of the animal rights movement*. Temple University Press, Philadelphia.

Francione, G. L. 2000. *Introduction to animal rights: Your child or the dog?* Temple University Press, Philadelphia.

Francione, G. L. and Charlton, A. E. 1992. *Vivisection and dissection in the classroom: A guide to conscientious objection*. The American Anti-Vivisection Society, Jenkintown, Pennsylvania.

Gentle, M. J. 1992. Pain in birds. *Animal Welfare* 1, 237-247.

Gibbons, E. F., Wyers, E. J, Waters, E., and Menzel, E. W. (editors) 1994. *Naturalistic environments in captivity for animal behavior research*. State University of New York Press, Albany, New York.

Gibbons, E. F., Jr., Durrant, B. S., and Demarest, J. (editors). 1995. Conservation of endangered species in captivity. State University of New York Press, Albany.

Godlovitch, S. and R. and Harris, J. (editors) 1972. *Animals, men and morals*. Taplinger, New York.

Gold, M. 1995. *Animal rights: Extending the circle of compassion*. Jon Carpenter, Oxford, England.

Goodall, J. 1990. *Through a window: My thirty years with the chimpanzees of Gombe*. Houghton Mifflin Company, Boston.

Goodall, J. 1994. Digging up the roots. *Orion* 13: 20-21.

Goodall, J. with Berman, P. 1999. *Reason for hope: A spiritual journey*. Warner Books, New York.

Goodman, B. 1991. Keeping anglers happy has a price: Ecological and genetic effects of stocking fish. *BioScience* 41: 294-299.

Goude, A. (editor) 1994. *The human impact on the natural environment*. MIT Press, Cambridge, Massachusetts.

Greek, C.R. and Greek, J.S. 2000. *Sacred cows and golden geese: the human cost of experiments on animals*. Continuum, New York.

Green, A. 1999. Animal underworld: Inside America's market for rare and exotic species. *PublicAffairs*, New York.

Guillermo, K. S. 1993. *Monkey business: The disturbing case that launched the animal rights movement*. National Press Books, Washington, D. C.

Hart, L. (editor) 1998. *Responsible conduct of research in animal behavior*. Oxford University Press, New York.

Hoage, R. J. (editor) 1989. *Perceptions of animals in American culture*. Smithsonian Institution Press, Washington D. C.

Johnson, A. 1992. *Factory farming*. Basil Blackwell.

Johnson, L. E. 1991. *A morally deep world: An essay on moral significance and environmental ethics*. Cambridge University Press, New York.

Kahn, P. H., Jr. 1999. *The human relationship with nature: Development and culture*. MIT Press, Cambridge, Massachusetts.

Kellert, S. R. and Wilson, E. O. (editors). 1993. *The biophilia hypothesis*. Island Press, Washington, D. C.

Kew, B. 1991. *The pocketbook of animal facts and figures*. Green Print, London.

Kirkwood, J. 1992. Wild animal welfare. In Ryder, R. D. and Singer. P. (editors) *Animal welfare and the environment*. Duckworth, London. pp. 139-154.

Kistler, J. (editor). 2000. *Animal rights: A subject guide, bibliography, and internet companion*. Greenwood Publishing Group, Westport, Connecticut.

Kleiman, D. G., Allen, M. E., Thompson, K. V., and Lumpkin, S. (editors) 1996. *Wild mammals in captivity: Principles and techniques*. University of Chicago Press, Chicago.

Knight, R. L. and Gutzwiller, K. J. (editor) 1995. *Wildlife and recreationists: Coexistence through management and research*. Island Press, Washington, D. C.

Krause, M. 1996. Biological continuity and great ape rights. *Animal Law* 2, 171-178.

LaFollette, H. and Shanks, N. 1996. *Brute science: The Dilemmas of animal experimentation*. Routledge, New York.

Linzey, A. 1976. *Animal rights*. SCM Press, London.

Lockwood, R. and Ascione, F. R. (editors) 1998. *Cruelty to animals and interpersonal violence: Readings in research and application*. Purdue University Press, West Lafayette, Indiana.

Lyman, H. F. 1998. *Mad cowboy: Plain truth from the cattle rancher who won't eat meat*. Scribner, New York.

Mack, A. (editor) 1999. *Humans and other animals*. Ohio State University Press, Columbus.

Magel, C. R. 1989. *Keyguide to information sources in animal rights*. McFarland, Jefferson, North Carolina.

Manes, C. 1997. *Other creations: Rediscovering the spirituality of animals*. Doubleday, New York.

Manning, A. and Serpell, J. (editors) 1994. *Animals and human society: Changing perspectives*. Routledge, New York.

Martin, Ann, N. 1997. *Foods pets die for: Shocking facts about pet food*. NewSage Press, Troutdale, Oregon.

Mason, J. 1993. *An unnatural order: Uncovering the roots of our domination of nature and each other*. Simon & Schuster, New York.

Mason, J. and Singer, P. 1980. *Animal factories*. Crown, New York.

Masson, J. M. and McCarthy, S. 1995. *When elephants weep: The emotional lives of animals*. Delacourte Press, New York.

Medical Research Modernization Committee. 1998. A critical look at animal experimentation. New York, New York.

Meyers, G. 1998. *Children and animals: Social development and our connections to other species*. Westview Press, Boulder, Colorado.

Midgley, M. 1983. *Animals and why they matter*. University of Georgia Press, Athens.

Mighetto, L. 1991. *Wild animals and American environmental ethics*. University of Arizona Press, Tucson.

Newkirk, I. 1990. *Save the animals: 101 easy things you can do*. Warner Books, New York.

Nichols, M. and Goodall, J. 1999. *Brutal kinship*. Aperture, New York.

O'Barry, R. with Coulbourn, K. 1988. *Behind the Dolphin Smile*. Algonquin Books, Chapel Hill, North Carolina.

Orion Society. 1995. *Bringing the world alive: A bibliography of nature stories for children*. Orion Society, Great Barrington, Massachusetts.

Orlans, F. B. 1993. *In the name of science: Issues in responsible animal experimentation.* Oxford University Press, New York.

Orlans, F. B., Beauchamp, T. L., Dresser, R., Morton, D. B., and Gluck, J. P. (editors) 1998. *The human use of animals: Case studies in ethical choice.* Oxford University Press, New York.

Paul, E. S. 1996. The representation of animals on children's television. *Anthrozoös* 9: 169-181.

Peterson, D. and Goodall, J. 1993. *Visions of caliban: On chimpanzees and people.* Houghton Mifflin Company, Boston.

Pluhar, E. B. 1995. *Beyond prejudice: The moral significance of human and non-human animals.* Duke University Press, Durham, North Carolina.

Poole, J. 1998. An exploration of a commonality between ourselves and elephants. *Etica & Animali.* 9/98, 85-110.

Quinn, D. 1992. *Ishmael.* Bantam, New York.

Rachels, J. 1990. *Created from animals: The moral implications of Darwinism.* Oxford University Press, New York.

Regan, T. 1983. *The case for animal rights.* University of California Press, Berkeley.

Regan, T. and Singer, P. (editors) 1989. *Animal rights and human obligations.* second edition. Prentice-Hall, New Jersey.

Regenstein, L. G. 1991. *Replenish the earth: A history of organized religion's treatment of animals and nature – including the Bible's message of conservation and kindness toward animals.* Crossroad, New York.

Rollin, B. E. 1981/1992. *Animal rights and human morality.* Prometheus Books, Buffalo, New York.

Rollin, B. E. 1989 *The unheeded cry: Animal consciousness, animal pain and science.* Oxford University Press, New York. (reprinted 1998, Iowa State University Press).

Rollin, B. E. 1995. *The Frankenstein syndrome: Ethical and social issues in the genetic engineering of animals.* Cambridge University Press, New York.

Rolston, H. III. 1989. *Environmental ethics: Duties to and values in the natural world.* Temple University Press, Philadelphia.

Rowe, M. (editor) 1999. *The way of compassion: Survival strategies for a world in crisis.* Stealth Technologies, Inc., New York.

Rowan, A. 1984. *Of mice, models, and men: A critical evaluation of animal research.* State University of New York Press, Albany.

Rowan, A. (editor) 1988. *Animals and people sharing the world.* University Press of New England, Hanover, New Hampshire.

Rowan, A. N., Loew, F. M., and Weer, J. C. 1995. *The animal research controversy: Protest, process & public policy – an analysis of strategic issues.* Tufts University School of Veterinary Medicine, Boston.

Russell, W. M. S. and Burch, R. L. 1959/1992. *The principles of humane experimental technique.* UFAW, Wheathampstead, England.

Ryder, R. D. 1989. *Animal revolution: Changing attitudes towards speciesism.* Blackwell, London.

Ryder, R. D. (1998). *The political animal: The conquest of speciesism.* McFarland & Company, Inc., Publishers. Jefferson, North Carolina and London.

Sapontzis, S. 1995. We should not allow dissection of animals. *Journal of Agricultural and Environmental Ethics* 8, 181-189.

Savage-Rumbaugh. E. S. 1997. Why are we afraid of apes with language? In A. B. Scheibel and J. W. Schopf (editors), *Origin and evolution of intelligence*. Jones and Bartlett, Sudbury, Massachusetts. pp. 43-69.

Schaller, G. B. 1993. *The last panda*. University of Chicago Press, Chicago.

Singer, P. (editor) 1986. *In defense of animals*. Harper and Row, New York.

Singer, P. 1990 *Animal liberation*, second edition. New York Review of Books, New York.

Singer, P. 1998. *Ethics into action: Henry Spira and the animal rights movement*. Rowman & Littlefield, Lanham, Maryland.

Shapiro, K. 1998. *Animal models of human psychology: Critique of science, ethics and policy*. Hogrefe & Huber Publishers, Seattle, Washington.

Shepherdson, D. J., J. D. Mellen, and M. Hutchins. (Eds.) 1998. *Second nature: Environmental enrichment for captive animals*. Smithsonian Institution Press, Washington, D. C.

Smith, D. A critique of the American Medical Association's white paper: Use of animals in biomedical research - the challenge and the response. American Anti-Vivisection Society, Jenkintown, Pennsylvania.

Sobel, D. 1996. *Beyond ecophobia: Reclaiming the heart in nature education*. Orion Society, Great Barrington, Massachusetts.

Sowing seeds: A humane education workbook. 1995. Animalearn and LivingEarth Learning Project. American Anti-Vivisection Society, Jenkintown, Pennsylvania.

Spira, H. 1986. Fighting to win. In Singer 1986, pages 194-208.

Stephens, M. L. 1986. *Maternal deprivation experiments in psychology: A critique of animal models*. American Anti-Vivisection Society, Jenkintown, Pennsylvania.

Tannenbaum J. 1995. *Veterinary ethics*. (2nd ed.) Mosby, St. Louis.

Taylor, P. W. 1986. *Respect for nature: A theory of environmental ethics*. Princeton University Press, Princeton, New Jersey.

Thanki, D. 1998. Virtual surgery in veterinary medicine. Animal Welfare Information Center, 9 (no. 1-2), 11

The animal rights handbook: Everyday ways to save animal lives. 1990. Living Planet Press, Los Angeles, California.

Tobias, M. 1998. *Nature's keepers: On the front lines of the fight to save wildlife in America*. John Wiley, New York.

Tobias, M. and Solisti, K. (editors) 1998. *Kinship with the animals*. Beyond Words Publishers, Portland, Oregon.

Tobias, M. 1999. *Voices from the underground: For the love of animals*. Hope Publishing House, Pasadena, California.

Wilcove, D. S., Rothstein, D., Dubow, J., Phillips, A., and Losos, E. 1998. Quantifying threats to imperiled species in the United States. *BioScience* 48: 607-615.

Wilkie, D. S. 1999. Bushmeat hunting in the Congo Basin: An assessment of impacts and options for mitigation. *Biodiversity and Conservation* 8, 927-955.

Wise, S.M. 2000. *Rattling the cage: Toward legal rights for animals*. Perseus Book, Cambridge, Massachusetts.

Woodroffe, R., Ginsberg, J., and Macdonald, D. (editors) 1997. *The African Wild Dog*. International Union for Conservation of Nature and Natural Resources, Gland, Switzerland.

Wynne-Tyson, J. (editor) 1988. *The extended circle: A commonplace book of animal rights.* Paragon House, New York.

Zimmerman M E, Callicott J B, Sessions G, Warren K J, and Clark J (editors) 1993. *Environmental philosophy: From animal rights to radical ecology.* Prentice-Hall, New York.

Zinko, U., Jukes, N., Gericke, C. 1997. *From guinea pig to computer mouse: Alternative methods for a humane education.* EuroNiche.

Zurlo, J., Rudacille, D., and Goldberg, A. M. 1994. *Animals and alternatives testing: History, science, and ethics.* Mary Ann Liebert Inc. Publishers, New York.

Journals and magazines

Agriculture and Human Values

Alternatives to Laboratory Animals

Animal Activist Alert

Animal Behaviour

Animal Biotechnology

Animal Issues

Animal Law

Animal Law Journal

Animal People

Animal Welfare

Animal Welfare Information Center Newsletter

Anthrozoös: A multidisciplinary journal of the interactions
of people and animals

Applied Animal Behaviour Science

AV Magazine, published by the American Anti-Vivisection Society

Between the Species: A Journal of Ethics

Biodiversity and Conservation

Biological Conservation

Biology and Philosophy

British Poultry Science

Bunny Huggers' Gazette

Conservation Biology

Directions (Association of Veterinarians for Animal Rights)

E The Environmental Magazine

Environmental Conservation

Environmental Ethics

Environmental Law

Environmental Values

Ethics and Behavior

Ethology

Etica & Animali

Hastings Center Report

Humane Society of the United States (HSUS) News

InterActions Bibliography (now, Humans & Other Species)

International Society for Environmental Ethics

Johns Hopkins Center for Alternatives to Animal Testing Newsletter

Journal of Agricultural and Environmental Ethics

Journal of the American Veterinary Medical Association

Journal of Animal Science

Journal of Applied Animal Research
Journal of Applied Animal Welfare Science
Journal of Zoo and Wildlife Medicine
Lab Animal
Laboratory Animals
Laboratory Animal Science
Laboratory Primate Newsletter
Medical Research Modernization Committees Report
Planet 2000 Newsletter
Royal Society for the Prevention of Cruelty to Animals Science Journal
Satya
Shelter Sense (now Animal Sheltering)
Society and Animals
The Animal Policy Report (Tufts University)
The Animals' Agenda
Wards
Wildlife Society Bulletin
Zoo Biology